Statistical Process Control
for Health Care

Marilyn K. Hart

Robert F. Hart

University of Wisconsin, Oshkosh

DUXBURY

THOMSON LEARNING

Australia • Canada • Mexico • Singapore • Spain • United Kingdom • United States

DUXBURY

✦ ™

THOMSON LEARNING

Sponsoring Editor: *Carolyn Crockett*
Marketing: *Tom Ziolkowski/Mona Weltmer*
Editorial Assistant: *Jennifer Jenkins*
Production: *Tessa Avila*
Permissions: *Sue Ewing*
Cover Design: *Denise Davidson*

Cover Photo: *Andrew Faulkner*
Print Buyer: *Kris Waller*
Cover Printing: *Phoenix Color Corp.*
Printing and Binding: *Maple-Vail Book Manufacturing Group*

For more information about this or any other Duxbury products, contact:
DUXBURY
511 Forest Lodge Road
Pacific Grove, CA 93950 USA
www.duxbury.com
1-800-423-0563 (Thomson Learning Academic Resource Center)

For permission to use material from this work, contact us by
www.thomsonrights.com
fax: 1-800-730-2215
phone: 1-800-730-2214

Printed in the United States of America

10   9   8   7   6   5   4   3   2   1

MINITAB is a trademark of Minitab, Inc., and is used herein with the owner's permission. Portions of MINITAB Statistical Software input and output contained in this book are printed with permission of Minitab, Inc. Statit is a registered trademark of Statware, Inc.

Library of Congress Cataloging-in-Publication Data

Hart, Marilyn K.
    Statistical process control for health care / Marilyn K. Hart, Robert F. Hart.
        p. cm.
    Includes index.
    ISBN 0-534-37865-X
    1. Medical care—Quality control—Statistical methods. 2. Health facilities—Evaluation—Statistical methods. I. Hart, Robert F. II. Title.

RA399.A1 H37 2001
362.1′068′5—dc21

Table of Contents

Chapter 4 Control Chart Theory and the I Chart for Time-Ordered Data   57

Chapter 5 The Xbar and s Chart   95

Foreword
by Carl D. Stevens, M.D., MPH
Clinical Associate Professor of Medicine
Los Angeles County Harbor – UCLA Medical Center

In keeping with modern, "just-in-time" production principles, this book arrives at precisely the moment when it is most needed to help spur the transition to a safer, more reliable healthcare system. In March 2001, with the publication of "Crossing the Quality Chasm," the Institute of Medicine set a new direction and issued a major challenge to providers of medical care in the United States, calling for a fundamental restructuring of the system to improve care. Following closely behind that publication, the Harts' book offers help with what for many organizations presents the greatest hurdle to meeting this challenge: learning to incorporate quantitative performance measurement into daily work routines as the foundation for an improvement-oriented culture. For health system managers across the nation who can see clearly where their organizations need to go but are uncertain about how to get there, the approach and methods taught in this book provide an important part of the answer.

Familiarity with control charts and other statistical process control methods has grown steadily since the mantra of continuous improvement spread from the manufacturing sector to health care just over a decade ago. However, the number of delivery systems that have successfully integrated SPC methods into their daily operations, and passed the resulting outcome and cost improvements on to their patients, remains quite small. Several factors have contributed to the slow adoption of these methods. One is the scarcity of high-quality, clinically detailed data to support valid performance measurement. A second barrier has been the persistence of a "silo" mentality in many healthcare organizations, which separates the clinical and financial management functions. Both of these barriers are beginning to fall, however, as clinical information systems improve data availability and market conditions increasingly dictate full integration of clinical and financial management. Along with a stronger focus on customers and teamwork, performance measurement will emerge as a key success strategy for healthcare delivery systems over the coming decade. SPC will rapidly spread beyond the current group of leading-edge centers to become a standard feature of healthcare operations.

To fulfill this promise, our nation will need to quickly train a cadre of healthcare practitioners and data analysts in the philosophy of continuous improvement and the quantitative methods of SPC. This book can play a major role in helping individuals and organizations meet this objective. The book follows the training sequence used by the authors in their highly acclaimed training seminars, which have provided physicians, nurses, and quality managers from across the United States with the basic skills they need to carry out quality improvement projects using SPC methods. From its beginning, the book takes a hands-on approach, placing emphasis on getting started on projects quickly and getting results. The examples and case studies are drawn from real improvement projects undertaken by actual hospitals and medical groups. In addition to explaining the selection of charts and the underlying calculations, the authors have included sections on the "management considerations" that lay out in detail how decision makers can best take advantage of SPC tools to explore opportunities and improve care. In the course of their seminars, the Harts have fielded dozens of questions on "How do I apply this method to my data sets, and how do I use the results to improve outcomes at my institution?" Their answers to many of these questions are included in the text.

Along with their extensive teaching experience, the authors bring other unique strengths to their enterprise. The Harts were students of W. Edwards Deming and are extremely effective at communicating the philosophy, as well as the methodology, that makes SPC such a compelling and appealing management technique. The emphasis on teamwork, the open-minded search for solutions, and learning to hear objectively what the data are saying permeates the book. In the course of their research and teaching activities, the Harts have developed personal relationships with the senior medical management teams at a number of leading-edge delivery systems, and the book reflects their resulting real-world perspective on what works and what doesn't in adapting SPC to the health care setting.

Even with the help of excellent books like this one, learning to use SPC methods effectively and accurately presents a challenge. Mastery of these techniques requires active engagement, frequent practice, and—as the Harts delight in pointing out—the use of simple arithmetic, which continues to scare the daylights out of many of us. Readers will return to the text frequently as they gain familiarity with the different control chart procedures and will build confidence in presenting their analyses to others. In most organizations, the SPC expert wears many hats and contributes to many aspects of the quality effort, including selection of projects and measures, study design, data collection, analysis, and championing the continuous improvement philosophy. By sticking to basics and emphasizing the doable, this book will serve as a valuable resource to those who agree to take on this challenge on behalf of their organizations. When the "internal expert," regardless of experience, begins to receive more requests for help than he or she can handle, the organization is on its way. When the executive team refuses to start a meeting without the latest control charts, it has arrived.

Measurement lies at the heart of genuine quality improvement, the kind that healthcare organizations undertake on behalf of their patients and communities, not simply to ensure accreditation. When delivery systems get ready to transition from talking about continuous quality improvement to really practicing it, learning to measure and manage care processes and outcomes becomes the first priority. If quality is Job One, measurement is Job Zero. This book will help organizations start down the road to the future.

Preface

Dr. Walter A. Shewhart invented the control chart as a tool for quality improvement in the 1920s, publishing his groundbreaking book *Economic Control of Quality of Manufactured Product* in 1931. Much excellent work has been done since then by some great authors: Western Electric (now AT&T), 1956; Grant and Leavenworth, originally 1946, 1996; and Duncan, originally 1952, 1986, to name a few. Unfortunately, much of the original work has been forgotten. In order to learn from these gems, whenever possible we have quoted from these original works. This is also in the hope that the reader may turn to these works to learn from the masters.

In order to take the "black box" out of the computer-generated solutions, we have attempted to explain the statistics in basic terms and illustrate the calculations with very short examples. In reality, these examples do not contain enough data to have any faith in the statistical results. They only serve to clarify the calculations. These short examples may be skipped, if desired, and emphasis put only on application and interpretation.

Real cases use much more data than were used in these short examples. (Guidelines for quantities of data needed are given in the various chapters.) For the Computer Supplement sections, we have chosen to use Statit Express QC by Statware, Inc., Corvallis, Oregon, www.statware.com, and Minitab Student Version by Minitab Inc., State College, Pennsylvania, www.minitab.com. Both are technically correct (no small matter since some statistical packages are incorrect). Statit is superior for control charts; Minitab is easier for keying in data and making histograms. The graphs in this book were made by Statit Professional.

These statistical process control (SPC) methods are often associated with quality control because this is where they were developed and originally used. However, we feel very strongly that these SPC methods are not just "quality control" techniques. They are really an alternative statistical approach to classical statistical methods such as hypothesis testing. Control charts can be used to monitor data over time, but also to compare data from different "populations," such as for comparing laboratory turnaround times from two different shifts.

The case studies are all from real data sets obtained from many of our friends and colleagues. They have been modified to protect confidentiality. Note that actual situations are not as simple as textbook examples; they contain the complexities of the "real world." It is hoped that while working through these case studies the reader will become comfortable with looking at data several different ways to obtain information.

The authors gratefully acknowledge the help and ideas given by Tony Abe, Marilyn Bufton, Marilyn Lord, Lorie Mitchell, Elizabeth Rhodes, Larry Staker, Richard Stanula, and Laurie Sullivan. Special thanks are due to Ray Carey and Marc Pierson for their substantial contributions and to Adam Jansen for his editing help. In addition, the kind help and support of Cal Bonine, Mathieu Federspiel, Guy March, Tom Simas, and Jamie Wyant from Statware, Inc., have been much appreciated.

Chapter 1 Objectives and Preliminary Information

Health care is on the minds of most Americans. Many changes have been made in the past decade to reduce costs and increase quality. However, there are still many opportunities for improvement, primarily due to the poor use of data to continuously improve the quality of care.

In general, industry has made better use of their data. Statistical process control (SPC), the data analysis method of quality improvement, has been in use since the 1930s. In the last 20 years, primarily under the influence of the automotive industry, the use of data to analyze processes and thereby find ways to improve them has undergone much standardization.

The fact that health care needs to learn from industry is widely recognized. Dennis O'Leary, president of the Joint Commission on Accreditation for Healthcare Organizations (JCAHO), states, "We need to recognize that the most useful information is provided by logical groupings of measures, not individual measures. That reality is well established in the industrial quality improvement model" [1998, p. 39]. Since healthcare organizations are just beginning to use SPC for quality improvement, no standard procedures exist. In many healthcare organizations, no SPC is used at all. For instance, as one physician laments in his article "Physician, Measure Thyself" [Newcomer, 1998, p. 32]:

> I can not tell you if I am an excellent oncologist or merely a mediocre one. The answer to that question is locked in the medical records room. . . . Most physicians have no performance measurements to assess their clinical competency. They use highly subjective methods such as peer opinion, referral rates, or self-evaluation of anecdotal incidents. . . . All of these methods lack the unbiased truthfulness of measured data, but objective practice data are rarely available. Unfortunately, the only valid and reliable data I received about my practice were monthly financial billing summaries.

The prologue to David Eddy's article [1998, p. 7] states that "the available tools to accurately measure performance of providers are quite limited." Further, in 1999, JCAHO required the following [1999, PI 4.1]:

> Appropriate statistical techniques are used to analyze and display data. . . . Some of the types of statistical tools that could be considered are
> • run charts, which display summary and comparative data;
> • control charts, which display variation and trends over time;
> • histograms;
> • Pareto charts;
> • cause-and-effect or fishbone diagrams; and
> • other statistical tools, as appropriate.

Objectives

This text has three objectives:

1. Understand the theory of *statistical process control* (SPC*)*.

It is relatively easy to collect data. Piles (or gigabytes) of it often exist. It is more difficult to extract information from the data. SPC is the branch of statistics that uses basic statistical techniques (requiring only simple math calculations) to organize the data so that the information becomes obvious. (*Note*: In this book, very short, easy examples are used to illustrate the mathematical calculations involved in order to take the "black box" approach out of the analyses. However, the hand calculations may be skipped if desired.)

The first objective of the text is to answer these questions: *What should be done with the data?* and *Why?*

2. Analyze the data on the computer.

Real-life data sets are large, and hence, very time consuming to analyze with hand calculations. Therefore, it becomes imperative to learn how to do these analyses using the computer. The supplements to the chapters use Statit Express QC (www.statware.com) and Minitab Student Version (www.minitab.com). These software packages are easy to use, handle all of the statistical techniques used in this text, and are used by some healthcare professionals. However, the data sets used in this text are also supplied in Excel format so that any statistical package may be used. These supplements will answer these questions: *How do you input the data*? and *How are the charts made?*

3. Apply these techniques to real data.

It is important to understand SPC beyond the simple textbook examples and to get into the complexity of actual experience. Hence, the case studies have been adapted from real data sets, modified for confidentiality reasons. The case studies attempt to answer these questions: *Which techniques are appropriate?, What are the charts telling us?,* and *How does this help with quality improvement?*

Understanding Variation

There is variation in everything. Dr. Walter A. Shewhart [1931], the inventor of the control chart and SPC, pointed out that if the letter "a" is written several times, the letters will not look identical. Yet the inputs to the process appeared to be the same—the same person, method, materials, environment, and measurement system (one's eyes). The reality is that no matter how alike the inputs to the process are, the outputs will vary.

As an example of process variation, suppose you decide to roll two dice many times and report back the sum of the two dice. The lowest you expect to roll is a sum of two, the highest is a sum of twelve. You would expect to average seven. If your first roll was a five, this would not be considered unusual. That is simply the nature of rolling two dice. However, if you rolled a one you would look to see what had happened. (Perhaps you dropped one of the dice on the floor.) You would find out what happened and take corrective action.

Once you become careful so you no longer drop the dice, the sum becomes predictable—always between two and twelve. Shewhart called this predictability *control* [1931, p. 6]:

> ... a phenomenon will be said to be controlled when, through the use of past experience, we can predict, at least within limits, how the phenomenon may be expected to vary in the future. Here it is understood that prediction within limits means that we can state, at least approximately, the probability that the observed phenomenon will fall within the given limits.

The variability that appears to be inherent in a controlled process is due to what Shewhart called *chance causes* [1931, p. 7]:

> In all forms of prediction an element of chance enters. The specific problem that concerns us at the present moment is the formulation of a scientific basis for prediction, taking into account the element of chance, where, for the purpose of our discussion, *any unknown cause of a phenomenon will be termed a chance cause.*

Dr. W. Edwards Deming, as a world-renowned quality improvement expert, was Shewhart's most famous student. Deming called these random causes *common causes* [1975, p. 3]:

> *Faults of the system (common or environmental causes)* . . . These faults stay in the system until reduced by management. Their combined effect is usually easy to measure. . . . Common causes get their name from the fact that they are common to a whole group of workers: they belong to the system.

Shewhart [1931] introduced the terminology that a process with only common-cause variation was a controlled process and was in a *state of statistical control*. Pearson [1935] called such a process *statistically uniform*. A process that is statistically uniform has variation that is due only to random chance, the result of only common causes. The data are

> *homogeneous*
> *stable*
> *random*
> *without trends, spikes, steps, or cyclical patterns.*

In a controlled process, the variation is *not* due to assignable, special, or uncontrolled causes, and the data are free from systematic variation, trends, spikes, steps, or cyclical patterns.

Any variation of the sum of two dice outside of the limits of two and twelve would be attributed to *assignable causes* as Shewhart called them [1931] or *special causes* as Deming called them [1975]. When a special cause appears to occur, it is important to investigate. If the results are detrimental (such as the dropping of one of the dice), it must be corrected (picking it back up). However, it may be an improvement (maybe you like smaller sums). In that case, you would like to make it standard operating procedure (from now on, you will roll only one die).

Working with dice is nice because the limits on the sums are known—between two and twelve when rolling two dice. In real processes, there is no way to tell ahead of time what the variation due to common causes will be. That is why the statistical tool called a *control chart* is necessary. The control chart estimates the limits of the variation due to common causes. Any variation outside of those limits will indicate a special cause of variation. The task at hand will be to determine the nature of that special cause and either eliminate it (if it is detrimental) or make it standard procedure (if it is an improvement). At the very minimum, the special cause must be taken into account in the analysis.

A final note on this example is in order. Suppose you continue to roll two dice with sums between two and twelve, with only common causes present. However, further suppose that you don't want any sums above ten, perhaps because they are too expensive. Having only common causes present does not mean that you are necessarily happy with the process. It only means that the process is repeatable within limits, with only chance variation. The variation due to common causes may not yield acceptable process results. You cannot identify and eliminate special causes; there aren't any. The process itself must be changed. The variation due to common causes must be decreased. That is usually a more difficult task. Special causes are "unusual" happenings; common causes happen all the time—they are part of the usual process. Common causes must be identified through expert knowledge of the process and/or through experiments.

There are two fundamental uses for SPC and the control chart:

1.   As a road map for process improvement
2.   To judge whether a state of control may be inferred.

Process Improvement

It is of vital importance to note that you cannot improve all processes at once. Resources are limited and must be used judiciously. It is important to choose which process should be improved first and, in particular, which aspect of that process should be improved first. This decision is based on the subjective opinion of what is worth doing and what is doable.

The selection of the process to improve is based upon three criteria:

1.   What internal and/or external customers perceive to be quality problems or critical processes that need to be improved
2.   What noncustomers perceive to be quality problems
    –   Governmental and nongovernmental entities (e.g., JCAHO) that make evaluations
    –   Upper management
3.   Opportunities to lower costs without compromising quality

After selecting the process to be improved, the quantities to measure must be determined. Then the measurement technique must be established and a plan must be developed to collect the appropriate data.

For example, suppose that a hospital is determining where to put their improvement efforts. The steps to be taken would go something like the following:

1.   *Determine the process to be improved.* The quality improvement (QI) team has noted numerous complaints about laboratory turnaround times (TATs) and has decided that improvement efforts will be directed to that area.

2.   *Determine the particular aspect of that process to be improved.* The QI team tallied the complaints in TATs and found that most of them were related to complete blood counts (CBCs). So improvement efforts will be focused on CBC TAT.

3.   *Determine what is to be measured and how it is to be measured.* The QI team determined that TAT on CBCs will be defined as the time from when the blood is drawn until the results are entered into the computer.

4.   *Determine a plan to collect the appropriate data.* The laboratory and the QI team jointly developed a form to record pertinent information such as what time the blood is drawn, by whom, on what day of the week, and date. They determined who will fill out the form, who will tally and analyze the data, and at what intervals.

Once a process has been identified as needing improvement, the first step to improve that process is to establish the initial process capability (discussed in Chapter 6) as a *baseline.* Otherwise, there is no way to tell whether a change made in the process results in an improvement. The steps for process improvement may be simplified as follows:

1.   Plan for the improvements.
   - Document the present process by estimating the process capability to define the original process, that is, establish a baseline.
   - Identify the possible changes in the process to bring about an improvement.
   - Select the change most likely to bring about an improvement.
2.   Do the plan. Make the change (hopefully an improvement) in the process.
3.   Check the process.
   - Estimate the process capability on the *new* process.
   - Compare the new process capability to the baseline to see whether the change is actually an improvement.
4.   Act on the findings.
   - If the change did not result in an improvement, delete it.
   - If the change resulted in an improvement, make it standard procedure.
5.   Repeat the cycle for further improvements.

This fits in well with what is often called the *Plan-Do-Check-Act* cycle, the Plan-Do-Study-Act cycle, or the Deming cycle, after Dr. W. Edwards Deming.

Judging Whether a State of Control May Be Inferred

Rather than using SPC specifically for process improvement, the objective may be to judge whether a process is in control so that future performance may be predicted. This may be done for evaluation, forecasting, scheduling, and so on.

Quality Improvement Tools

When identifying opportunities for quality improvement or actions to make the improvements, there are several quality "tools" that may be very useful: tally sheet, Pareto chart, cause-and-effect diagram, flowchart, run chart, histogram, probability plot, control chart, affinity diagram, and so on.

Tally Sheet

The *tally sheet*, sometimes called a *check sheet*, is used to tally the counts of specific events, usually undesirable events. For instance, in order to avoid medication errors before they happen, a hospital decided to record incidents that potentially could have resulted in a medication error. A tally sheet was posted after being designed by a quality team of the people involved. As "Near Miss" situations were found, a tick mark was made on the appropriate error of the tally sheet in Figure 1.1.

Near Miss	Tally
Handwritten orders illegible	////
Incorrect dose ordered	///
Look-alike medications	//
Scheduled medication missed	////
Incorrect scheduling by RN	//
Medication unavailable	///// ///// ///// ///
Medication stocked in wrong place	///
Other errors	/

Figure 1.1 Tally Sheet

The tally sheet helps to clearly identify which events are greatest in frequency of occurrence. This helps direct efforts for quality improvement. In this case, efforts may be directed to improving quality by finding out why medications may not be available and how to remedy the situation.

Pareto Chart

The *Pareto chart* graphs the counts of various occurrences in decreasing order of frequency. Continuing with the example of occurrences of incidents of potential errors, a Pareto chart (Figure 1.2) is made on the results of the tally sheet in Figure 1.1. Note how the Pareto chart makes an informative display of the frequency of occurrences. It is very obvious from the Pareto chart that the unavailability of the proper medication is the most frequently occurring problem, again pointing to the fact that efforts may be directed to improving quality by finding out why medications may not be available and how to remedy the situation.

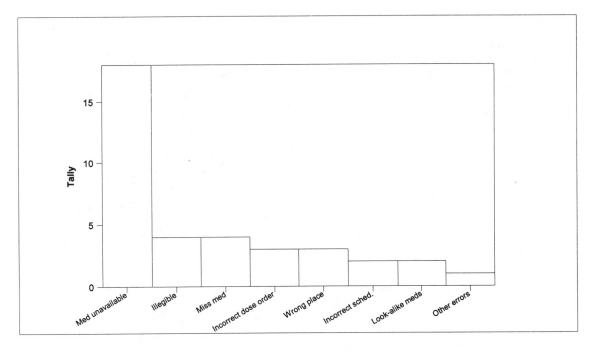

Figure 1.2 Pareto Chart

Joseph Juran named this technique after Vilfredo Pareto, an Italian economist who found that approximately 85% of the wealth was owned by 15% of the population. In fact, this "Pareto principle" is often called the "80-20 rule" or the "Law of the Significant Few" with 80% of the "problems" being caused by 20% of the types of occurrences. These 20% are called the *vital few*. In areas other than quality, this prioritizing by totals (or frequencies) is sometimes called "ABC Analysis" with the "A" items being the high-frequency ones.

Cause-and-Effect Diagram

The *cause-and-effect diagram* attempts to identify the inputs into a process since one or more of these inputs will be the cause of the problem. This will help with the solution of the problem. The cause-and-effect diagram graphs the inputs of materials, methods, machines, measurement, and people on a "fishbone" type of graph. Consequently, it is sometimes called a *fishbone diagram*. It is also known as an *Ishikawa diagram* after Kaoru Ishikawa, the man given credit for introducing it in Japan.

Continuing with the example of incidents of potential errors, a team of people involved in the situation brainstormed the inputs of the process that might result in medications not being available. An abbreviated cause-and-effect diagram is shown in Figure 1.3. The most likely cause (or causes) will be identified by the team and investigated further. Appropriate corrective action may then be taken. As noted above for the Plan-Do-Check-Act cycle, it is important to collect data before and after the corrective action is taken. It may be that the "corrective action" to the situation was misdirected and the situation became worse.

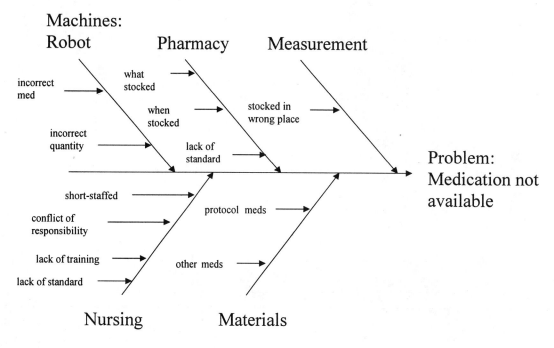

Figure 1.3 Cause-and-Effect Diagram

Flowchart

Often the reason that mistakes occur is that there is no well-defined process, just confusion. One use of the *flowchart*, a graphical portrayal of the steps of the process, is to define the steps of the present process. This is not an easy undertaking. A team of the people doing the task to be flowcharted must spell out exactly what the steps are, and often they discover that they are not all doing the same steps. Through some compromises, they must reach consensus on what the steps are. They may also use a flowchart by defining what the "best practice" steps should be and changing the process accordingly. For example, a team of care providers may research best practices and use their own knowledge and experience to prepare a draft protocol (in written or flowchart form). This draft is distributed to all physicians, nurses, and others who will use it in order to obtain their input. Using their feedback, the protocol is modified in an iterative process until a consensus is reached. This protocol is then tried on a few patients and further modified through more iterative steps [Clemmer, et al, 1999]. (In industry, this final form has sometimes been called a "current best approach" admitting that further changes may be necessary down the line.) Special situations with unique conditions may subsequently occur that justify deviations from the newly defined process. These must be documented and taken into account when the protocol is revised. An example of a simple flowchart for a patient monitoring his own blood pressure is found in Figure 1.4.

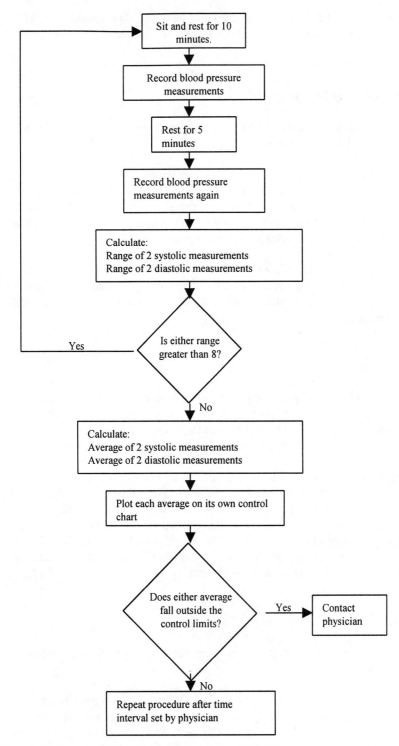

Figure 1.4 Flowchart for a Patient Monitoring Blood Pressure

As the flowchart becomes more complex, a *conceptual flowchart* may be useful. This consists only of the main steps (usually 3 to 12) of the process. Then each main step is expanded in more detail on its own flowchart.

Run Chart, Histogram, Probability Plot, and Control Chart

The run chart, histogram, probability plot, and control chart are used extensively throughout this book in the examples and case studies. They are briefly summarized here.

The run chart is a graph of the data in order of time. It looks for nonrandom patterns, trends, and points that are "obviously excessive." This helps identify special causes of variation. However, "obviously excessive" is not a statistical test but only a judgment call—and judgment can be misleading. The control chart is a statistical tool that will determine the limits of the common-cause system. Any points falling outside that system will signal special-cause variation that will need to be investigated.

The histogram is a graph that displays the variation in a set of data. It shows the frequency of each particular value (or range of values) of that data set and gives information about the population from which the data came. The probability plot also graphs the distribution of the data set, but gives more information about the population not found on the histogram.

Affinity Diagram

The affinity diagram (also known as a KJ diagram) is one of the "Seven Management and Planning Tools" [Brassard, 1996]. The affinity diagram is handy as a way of doing some quiet brainstorming. It is used to generate ideas and then groups these ideas based on natural relationships. For instance, you may be looking for the inputs for the cause-and-effect diagram for the problem of medications not being available when needed. After the issue has been well defined, ideas are generated in silence by each member of the team, with each idea briefly recorded on a separate card. When it appears that all members on the team have exhausted their ideas, the team leader suggests they begin to organize them on a desk, still in silence. Still without talking, the members randomly lay out the completed cards, then begin to sort the cards into related groupings. Cards may be moved back and forth until all team members quit moving them. Only then is talking allowed as the team members suggest heading cards for the various groupings. The result is called the *affinity diagram*. The advantages of the affinity diagram include the fact that ideas can be generated quickly and that even the quietest people on the team have equal input.

Other Quality Improvement Tools

There are many other quality improvement tools that are not the main thrust of this book, but some are worth mentioning here.

Poka-yoke was invented and developed by Shigeo Shingo as a technique for avoiding simple human and machine errors at work—a way of mistake-proofing the system. For instance, a weekly pill holder with a separate section for each day may be filled at the beginning of each week. Then throughout the week, the medications are taken each day and the holder empties. There can then be no confusion as to whether today's medication was taken or not.

Design of experiments is the discipline that designs an experiment in one of certain defined ways that allow for very straightforward statistical analysis. Their purpose is to discover the key variables that influence quality.

Regression, from classical statistics, is the statistical technique that calculates an equation that is used to estimate the relationships between two or more variables and measures the strength of that relationship.

Six Sigma, originally incorporated at Motorola, was initiated to increase customer satisfaction by recognizing customer needs and designing the process to meet those needs. It emphasizes mistake-proofing the process by simplification and standardization and by the institution of failure-free methodologies. Six Sigma then encourages continuous improvement of the process by measuring, analyzing, and controlling the improved process. It relies heavily on the use of SPC tools and problem solving. The goal of the Six Sigma philosophy is to decrease the error rate to less than 3.4 "mistakes" per million opportunities.

The *Malcolm Baldrige National Quality Award* was created in 1987 by an act of Congress (The Malcolm Baldrige Quality Improvement Act of 1987, Public Law 100-107). This law states that the award was created to improve quality and productivity by [National Institute of Science and Technology, www.quality.nist.gov]

a. helping to stimulate American companies to improve quality and productivity for the pride of recognition while obtaining a competitive edge through increased profits;
b. recognizing the achievements of those companies that improve the quality of their goods and services and providing an example to others;
c. establishing guidelines and criteria that can be used by other business, industrial, governmental, and other organizations in evaluating their own quality improvement efforts; and
d. providing specific guidance for other American organizations that wish to learn how to manage for high quality by making available detailed information on how winning organizations were able to change their cultures and achieve eminence.

The "*2001 Health Care Criteria for Performance Excellence*" for the Baldrige Award are organized into seven categories: leadership; strategic planning; focus on patients, other customers, and markets; information and analysis; staff focus; process management; and organizational performance results. The full text can be found at the web site of the National Institute of Standards and Technology (www.quality.nist.gov).

Types of Data

Two types of data are discussed in this book. Variables data are usually the result of measurements (such as time, weight, or volume) from a continuous scale. Hence, they are sometimes called a *continuous variable*. These are rounded to some increment (such as to the nearest day, nearest gram, or nearest cubic centimeter). Variables data are discussed in Chapters 2 through 6. Chapters 7 and 8 cover *attribute* data (also known as count-type data). Attribute data are the result of counting the number of occurrences of a characteristic, rather than measuring.

Chapter Summary

- There is variation in everything.

- Of primary concern is the separation of variation into special-cause variation and common-cause variation.

- A baseline must be established before changes are made in order to determine if changes are really improvements.

- The Plan-Do-Check-Act cycle leads to process improvement.

- There are many helpful quality improvement tools.

- Data may generally be separated into two types: variables and attribute.

Problems

For Problems 1 – 10, tell whether the data are variables or attribute data.

1. Length of stay subgrouped by diagnosis-related group (DRG).

2. Cost per case for each case in order of time.

3. Accounts receivable: days since billing subgrouped by payer.

4. Percent primary cesarean sections (subgrouped by physician, payer, month).

5. Number of cardiac rehab patient visits per week.

6. For each patient, the time is recorded from the time of the decision to admit the patient to the time of transfer to the unit.

7. Time from order entry to result.

8. Recorded each day: the number of reports that are more than three days old.

9. Percentage medication errors.

10. Number of critical care unit (CCU) infections per patient day subgrouped by month.

11. The process appears to have only random variation. It is the result of _____ causes.

12.	The process has some unusual outcomes exhibiting obviously excessive data points. This is probably due to _____ causes.

13.	Explain the Plan-Do-Check-Act cycle.

14.	Give an example of a problem area and how it could be approached by the Plan-Do-Check-Act cycle.

15.	Design a tally sheet for reasons for patient falls.

16.	Pick a problem area in health care and design a tally sheet for it.

17.	For the following tally sheet, make a Pareto chart.

Incorrect	Tally
Scheduling	///
Medication	///// ///// ///// ///// //
Route	///// ///
Time/Frequency	//
Concentration	/

18.	Expand on the cause-and-effect diagram in Figure 1.3.

19.	Make a cause-and-effect diagram for reasons for patient falls.

20.	Pick a problem area in health care and make a cause-and-effect diagram for it.

Chapter 2 Variables Data: Basics of Statistics and Graphs

As mentioned in Chapter 1, variables data are usually the result of measurements. These measurements are made upon quantities, which are continuous by nature (such as length or time) and, hence, are sometimes called *continuous variables*. Variables data also include percentages derived from measurements, such as the percentage of solids in a solution.

Population versus Sample

If you are to study the length of stay for patients after a laparoscopic cholecystectomy procedure (lap chole), the potentially infinite number of patients who might have a procedure in the future may be regarded as the *population*. Note that it is impossible to find out what the length of stay would be from all such patients, instead a subset of patients is studied. Such a subset is called a *sample* (or *subgroup*). It is hoped that the sample will be representative of the population. The number of patients studied, that is, the size of the sample, is symbolized by n.

Average

The measure most commonly used to describe a sample is a measure of its central tendency called the *mean* or *arithmetic average*. In this book this will be called simply the *average*. The average is calculated by adding the data in the sample and dividing by the number of observations in the sample. With variables data, a data point is symbolized by X and the average is symbolized by $\overline{X}$ or Xbar. In print it is often spelled out as Xbar and is read "X bar." The formula is

$$Xbar = \frac{\sum_{i=1}^{n} X_i}{n}$$

where n is the number of observations $X_1, X_2, \ldots, X_n$.

EXAMPLE 2.1

The length of stay (LOS) for a particular diagnosis-related group (DRG) appears to be increasing at a particular hospital, so it was decided that the LOS be studied. The LOS to the nearest day for each of the next five patients discharged with that DRG is recorded. The data are
    3, 5, 2, 8, 2
Note that $n = 5$ and Xbar = 20/5 = 4. □

EXAMPLE 2.2

The average LOS values for the DRG at three different hospitals are to be compared. The next seven patients discharged from each of the three hospitals are recorded below. Letting k be the number of subgroups, the LOS values are subgrouped by hospital, giving $k = 3$ subgroups of size $n = 7$.

Length of stay in days, subgrouped by hospital

	Hospital	
A	B	C
5	4	1
5	5	5
5	5	5
5	5	5
5	5	5
5	5	5
5	6	9

Note that for each hospital, $n = 7$ and Xbar = 5. How well did the average of 5 describe the data from hospital A? Very well, couldn't do better. How well did it describe the data from hospital B? Quite well. How well did it describe the data from hospital C? Not as well. Note that it is not enough to describe variables data with one number. A number to describe the variation is also needed. □

Measures of Variation: Range

As seen in Example 2.2, it is not sufficient to describe a set of variables data by its average. A measure must also be used to describe its variation. One measure of variation of the data is the range—the highest value minus the lowest value. The range is symbolized by R.

EXAMPLE 2.3

The LOS data in Example 2.1 was 3, 5, 2, 8, 2. The highest value is 8; the lowest is 2. The range is
    R = 8 - 2 = 6 □

EXAMPLE 2.4

Example 2.2 compared the average LOS from three hospitals. The ranges of the LOS values for the three hospitals are

Hospital	A	B	C
R =	0	2	8 □

EXAMPLE 2.5

Continuing with Example 2.2, comparing the average LOS from three different hospitals, a fourth hospital is added. As before, seven consecutive patients discharged from each of the hospitals are recorded below.

Length of stay in days, subgrouped by hospital

Hospital

A	B	C	D
5	4	1	1
5	5	5	1
5	5	5	1
5	5	5	5
5	5	5	9
5	5	5	9
5	6	9	9

Note that for each hospital, $n = 7$ and Xbar = 5. How well does an average of 5 describe the data from hospital D as compared to how it describes the data from hospital C? Not as well, since there appears to be even more variation in the LOS data from hospital D. However, the range of LOS values for hospital C and hospital D is each 8. Consequently, the range is not a good measure of variation. The range uses only two values from the data set—the highest and the lowest values. A measure of variation that uses all of the data is needed. □

Measures of Variation: Standard Deviation

The *standard deviation* is a measure of variation that uses all of the data. The estimate of the population standard deviation from a sample is symbolized by *s*, and it measures how much the individual observations differ from the average, Xbar.

$$s = \sqrt{\frac{\sum_{i=1}^{n}(X_i - Xbar)^2}{(n-1)}}$$

where *n* is the number of observations $X_1, X_2, \ldots, X_n$.

This is a mathematical formula that is handy for statisticians to use because it has many desirable properties. Many theorems and tables have been developed using this measure of variation.

The steps for calculating standard deviation of a sample are as follows:

1. Calculate the average, Xbar.
2. Calculate the deviation of each observation (X_i) from Xbar (i.e., $X_i - $ Xbar)
3. Square these numbers: $(X_i - $ Xbar$)^2$
4. Add these squares:

$$\sum_{i=1}^{n}(X_i - Xbar)^2$$

5. Divide by n - 1:

$$\frac{\sum_{i=1}^{n}(X_i - Xbar)^2}{(n-1)}$$

This is called the sample *variance*, symbolized by s^2. It is the unbiased estimate of the population variance.

6. Take the square root:

$$s = \sqrt{\frac{\sum_{i=1}^{n}(X_i - Xbar)^2}{(n-1)}}$$

This is the *standard deviation* symbolized by s.

EXAMPLE 2.6

Calculate the standard deviation for the data from hospital C in Example 2.5. Note that $n = 7$ and Xbar = 5 (step 1).

	Step 2 subtract average X_i - 5	Step 3 square
X_i	X_i - Xbar	$(X_i - \text{Xbar})^2$
1	-4	16
5	0	0
5	0	0
5	0	0
5	0	0
5	0	0
9	4	<u>16</u>

Step 4. Add: 32

Step 5. Divide by n - 1: $s^2 = 32/6 = 5.33$

Step 6. Take the square root:

$$s = \sqrt{5.33} = 2.31$$

Note: Arranging the data in order from the smallest to the largest value makes it easier to find the range, but it is not necessary for the computation of standard deviation. □

EXAMPLE 2.7

Compute the standard deviation for the data from hospital D in Example 2.5. Note that $n = 7$ and Xbar = 5 (step 1).

	Step 2 subtract average X_i - 5	Step 3 square
X_i	X_i - Xbar	$(X_i - \text{Xbar})^2$
1	-4	16
1	-4	16
1	-4	16
5	0	0
9	4	16
9	4	16
9	4	<u>16</u>

Step 4. Add: 96

Step 5. Divide by n - 1: $s^2 = 96/6 = 16$

Step 6. Take the square root:

$$s = \sqrt{16} = 4$$

Note that the standard deviation of Hospital D is larger than the standard deviation of Hospital C, reflecting the greater variation of data. □

EXAMPLE 2.8

The following data are the first 10 surgery times in minutes for laparoscopic cholecystectomy (lap chole) procedures from a larger data set (given later in Table 3.1) gathered over time.

110 120 100 90 105 130 75 95 105 103

Xbar = 1033/10 = 103.3

Largest measurement = 130, smallest measurement = 75, so R = 55

$n = 10$

X_i	X_i - Xbar	$(X_i - \text{Xbar})^2$
110	6.7	44.89
120	16.7	278.89
100	-3.3	10.89
90	-13.3	176.89
105	1.7	2.89
130	26.7	712.89
75	-28.3	800.89
95	-8.3	68.89
105	1.7	2.89
103	-0.3	0.09
		2100.1

Then $s^2 = 2100.1/9 = 233.34$

$$s = \sqrt{233.34} = 15.28 \quad □$$

An important purpose of sampling is to obtain estimates of the process average and standard deviation. However, these estimates are only meaningful if the process is stable and so defines what can be thought of as a "population." Control chart techniques for determining the stability of the process will be given in later chapters. The symbols given for the average and standard deviation of the process and sample are given in Table 2.1. Note that μ (mu) is the symbol for the "true" average of the process and σ (sigma) is the symbol for the "true" standard deviation of the process.

Table 2.1 Symbols for Average and Standard Deviation

	Average	Standard Deviation
Population or process	μ (mu)	σ (sigma)
Sample	Xbar	*s*

The Normal Distribution

If the graph of the frequencies (i.e., the histogram, discussed later in this chapter) of all the data appears to be symmetric and can be smoothed over with a bell-shaped curve (roughly like the one illustrated in Figure 2.1), it may be said to approximate a Gaussian or normal distribution. The normal distribution is easy to work with because much research has been done on it. In particular, if the true average (μ) and standard deviation (σ) of a normally distributed process are given, the complete distribution of all the data is known. Approximately 68% of the observations are within one standard deviation above or below the average, 95% are within two standard deviations, and 99.73% are within three standard deviations, as illustrated in Figure 2.2.

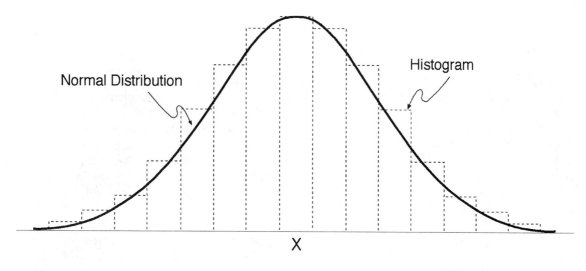

Figure 2.1 Normal Distribution with Histogram

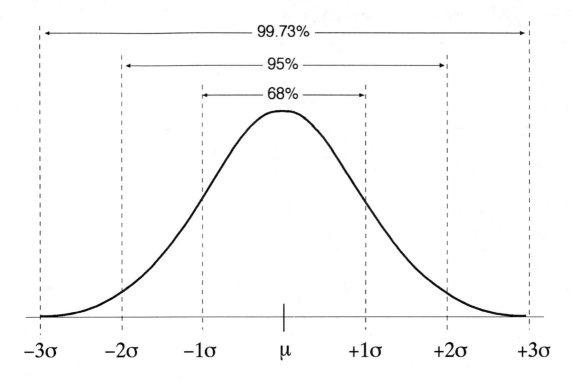

Figure 2.2 The Normal Distribution with Probabilities

EXAMPLE 2.9

If the lengths of stay for a particular DRG are normally distributed with an average of 8 days and a standard deviation of 2 days, approximately 68% of the stays will be within one standard deviation (2 days) of the 8-day average (i.e., between 6 and 10 days). About one-sixth of the population will fall in each tail (i.e., one in six will have a stay of less than 6 days and another one in six will have a stay of greater than 10 days). Approximately 95% will be within two standard deviations (4 days) of the 8-day average (i.e., between 4 and 12 days). About one in 40 will fall in each tail. One in 40 will have a stay of less than 4 days and another one in 40 will have a stay of greater than 12 days. Approximately 99.73% will be within three standard deviations (6 days) of the 8-day average (i.e., between 2 and 14 days). About one in 1,000 will have a stay of less than 2 days and one in 1,000 will have a stay of more than 14 days.

In reality, nothing is perfectly normally distributed, so these percentages would only be approximate. For most applications, the data only need be close to normal or *near-normal* for the techniques to be effective.□

Near-Normal Distributions

When using variables data, it is assumed that the process has a normal distribution. In fact, the process cannot have a normal distribution. The rounding introduced by measurements makes the real data discrete while the normal distribution is continuous. Further, the normal distribution goes from minus infinity to plus infinity, and real process data are always bounded. As Shapiro [1990, p. 5] warns,

> Any distribution, the normal for example, is a mathematical concept. Geary (1947) once suggested that in front of all statistical texts should be printed, "Normality is a myth. There never was and will never be, a normal distribution."

An important use for variables data is to make control charts to determine whether special-cause variation exists. A "near-normal" distribution is defined here as one where the departures from normality will have no detrimental effect on the utility of the control chart. The chart will still satisfactorily discriminate between common-cause and special-cause variation.

Histogram

The *histogram* is a graph of the frequency of each measurement from a data set. It is used to get a "quick picture" of the shape of the distribution of the data set. If the data take on only a few values with little variation between the low and high observations, the histogram will graph the frequency of each value. If the data take on many values, the data are usually grouped into *cells* or *bins*. When graphed, neighboring cells touch, reflecting the fact that the data are continuous. Where one cell ends, the next begins.

EXAMPLE 2.10

The following data are the first ten surgery times in minutes from laparoscopic cholecystectomies (lap choles) used in Example 2.8 from a larger data set gathered over time.

    110 120 100 90 105 130 75 95 105 103

The largest measurement is 130; the smallest measurement is 75. The range of the data is 55. If it is desired to make the cells of width 10, a logical way to define the cells would be

    71 – 80
    81 – 90
    91 – 100

and so on. The tally of the data is given in Table 2.2 and the histogram is given in Figure 2.3. □

Table 2.2 Tally of Data from Example 2.10 with Cell Widths of 10

Cell Numbers	Tally
71 to 80	/
81 to 90	/
91 to 100	//
101 to 110	////
111 to 120	/
121 to 130	/

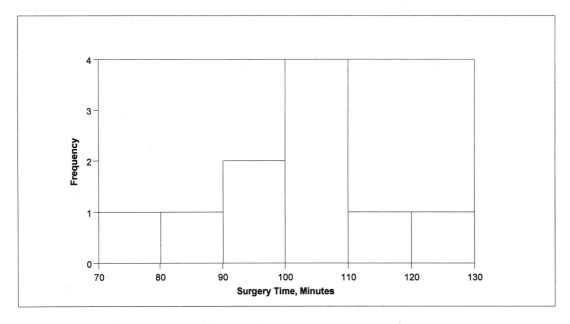

Figure 2.3 Histogram of Lap Chole Surgery Times (Cell Width = 10, n = 10)

EXAMPLE 2.11

If 12 cells are wanted, the cell width would be approximately 55/12 = 4.6, so a cell width of 5 would be used. The new tally is given in Table 2.3 and the histogram is shown in Figure 2.4. Note how the choice in cell width can give a different appearance to the shape of the distribution. Sometimes the difference can be drastic. □

Table 2.3 Tally of Data from Example 2.11 with Cell Widths of 5

Cell Numbers	Tally
71 to 75	/
76 to 80	
81 to 85	
86 to 90	/
91 to 95	/
96 to 100	/
101 to 105	///
106 to 110	/
111 to 115	
116 to 120	/
121 to 125	
126 to 130	/

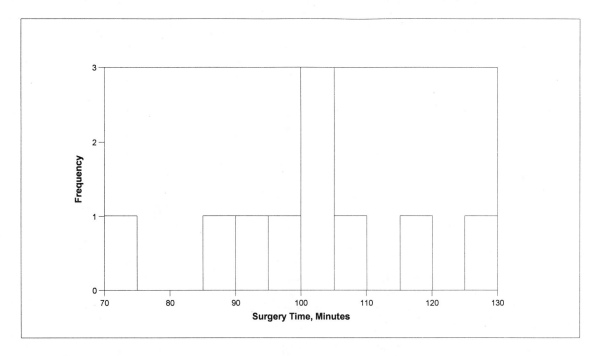

Figure 2.4 Histogram of Lap Chole Surgery Times (Cell Width = 5, n = 10)

You can get a rough idea of the shape of the distribution from the histogram. Figures 2.5, 2.6, and 2.7 are histograms for three synthesized data sets of 200 values each. Figure 2.5 is a histogram for 200 values from a normal distribution. Figure 2.6 is a histogram for 200 values from a distribution that is moderately skewed to the right (i.e., the histogram has a long tail toward the right, or high, end of the distribution). Figure 2.7 is a histogram for 200 values from a distribution that is much more severely skewed to the right. However, recall that the appearance of the histogram is somewhat dependent on the choice of cell width.

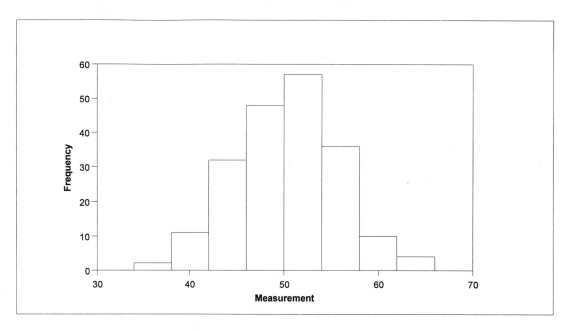

Figure 2.5 Histogram from a Synthesized Normal Distribution (*n* = 200)

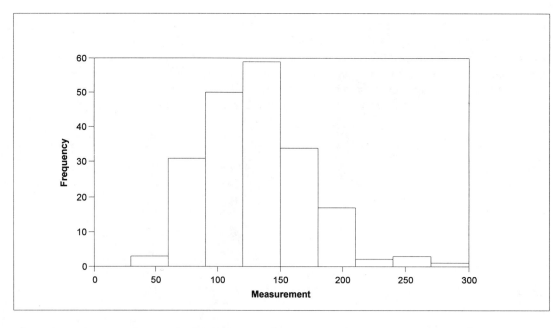

Figure 2.6 Histogram from a Distribution That Is Moderately Skewed to the Right (*n* = 200)

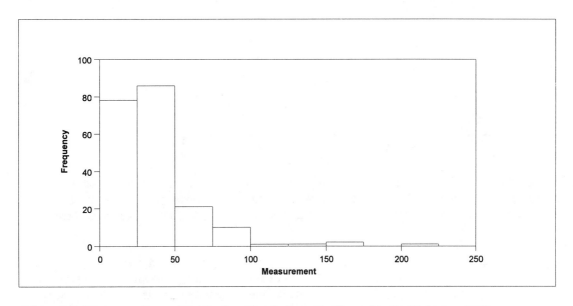

Figure 2.7 Histogram from a Distribution That Is Severely Skewed to the Right ($n = 200$)

Probability Plot

A probability plot is a graph of the relative cumulative frequencies of the data, using a specific plotting convention. A probability plot is used for two reasons:

1. It provides a graphical test to determine whether the data may be considered to be near-normal. It gives an excellent portrayal of the distribution of the data without the problem of being dependent on a choice of cell width as with the histogram.
2. It provides the best estimate for the percentages in the tails of the distribution and of the process capability (discussed in detail in Chapter 6).

To illustrate the calculations involved in making a probability plot, consider the very small data sample consisting of just three time-ordered values: 5, 3, and 9. In Table 2.4, the raw ordered observations of 3, 5, and 9 each occur only once, so each has a frequency of one. Note that they are arranged from the smallest observation (3) to the largest observation (9). The frequencies may then be accumulated, giving the *cumulative frequency* of the sample. Note that the cumulative frequency of 3 for the observation of 9 means that 3 observations from the sample are of value 9 or smaller. But this is 3 out of 3, 100% of the observations. This implies that 100% of the observations from the sample are of a value of 9 or smaller. The *relative cumulative frequency* also tells that 1 out of 3, 33%, of the observations from the sample are of value 3 or smaller and 2 out of 3, 67%, are of value 5 or smaller.

Table 2.4 Cumulative Frequency Calculations for the Sample ($n = 3$)

Data	Frequency	Cumulative Frequency	Relative Cumulative Frequency
3	1	1	1/3 = 0.33
5	1	2	2/3 = 0.67
9	1	3	3/3 = 1.00

Note that these cumulative frequencies are for the sample, but what is really desired are estimates of the relative cumulative frequencies (or cumulative percentages) for the total population (not just the sample). A *plotting convention* is needed for the population cumulative percentage, which will accomplish three objectives:

1. After putting data in ascending order, the middle value will be at 50%.
2. The largest observation *will not* be plotted at 100% but at some lower percentage. This will allow for some future observations from the process to be larger than the largest observation obtained from the sample.
3. The smallest observation will be plotted symmetrically to the largest (i.e., if the largest observation is plotted at x%, the smallest observation will be plotted at $(100 - x)$%).

There are many plotting conventions that accomplish these three objectives. One commonly used convention is $y = i/(n + 1)$, where y is the relative cumulative frequency, i is the cumulative frequency, and n is the total number of observations. Other plotting conventions that are sometimes used are

$$(i - 0.5)/n$$
$$(i - 3/8)/(n + 1/4)$$
$$(i - 0.3)/(n + 0.4).$$

The plotting convention of $i/(n + 1)$ will be used throughout this book. Its use is illustrated in Table 2.5.

Table 2.5 Probability Plot Calculations ($n = 3$)

Data	Frequency	Cumulative Frequency i	Relative Cumulative Frequency $i/(n + 1)$
3	1	1	1/4 = 0.25 = 25%
5	1	2	2/4 = 0.50 = 50%
9	1	3	3/4 = 0.75 = 75%

Notice that this plotting convention accomplishes the three objectives:

1. The middle value (5), called the *median*, is at 50%.
2. The largest value (9) is at 75%, leaving room for 25% of the population to be larger.
3. The smallest value (3) is at 25%, leaving room for 25% of the population to be smaller.

EXAMPLE 2.12

The first 10 lap chole surgery times used in Example 2.8 from a larger data set are used here as a short example to illustrate the hand calculations. Note that this very small data set is used only for a tutorial to illustrate the computational method.

Data: 110   120   100   90   105   130   75   95   105   103
Here, $n = 10$.

The ordered data are:
75   90   95   100   103   105   105   110   120   130

The probability plot calculations are given in Table 2.6. Since there are an even number of observations, the median is halfway between 103 and 105. □

Table 2.6 Probability Plot Calculations for Example 2.12

X	i	$i/(n+1) = i/11$
75	1	$1/11 = .09 = 9\%$
90	2	$2/11 = .18$
95	3	$3/11 = .27$
100	4	$4/11 = .36$
103	5	$5/11 = .45$
105	6	$6/11 = .55$
105	7	$7/11 = .64$
110	8	$8/11 = .73$
120	9	$9/11 = .82$
130	10	$10/11 = .91$

Making the Probability Plot

For the *probability plot* (also called the cumulative or normal probability plot), the relative cumulative frequencies are plotted on special graph paper (called normal probability paper). This paper has a special *y*-axis (vertical) scale that has been chosen so that data that are normally distributed will tend to yield a straight line. The *x*-axis (horizontal) scale is just a regular linear scale. Each ordered pair of the data point and its cumulative relative frequency are plotted.

EXAMPLE 2.13

Plotting the data from the ten values of Example 2.12, the lowest observation of 75 gets plotted at 9%, 90 gets plotted at 18%, and so on, as shown in Figure 2.8. Note that the plot of the data can reasonably be approximated by a straight line. ☐

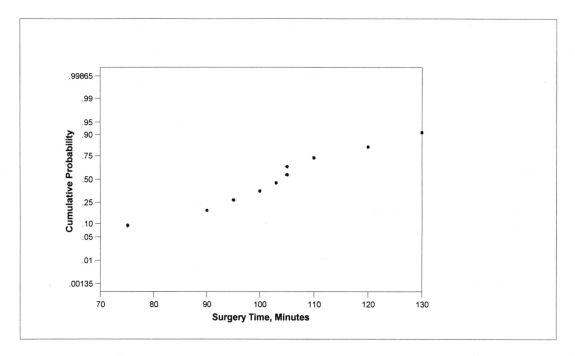

Figure 2.8 Probability Plot of Data in Example 2.13 ($n = 10$)

Shape of the Distribution

Figure 2.9 plots the 200 observations generated from the normal distribution that were displayed in the histogram in Figure 2.5. The plot of the data may be reasonably approximated by a straight line. Note that plotting 200 observations gives a much better feel for the shape of the distribution than does plotting only ten points.

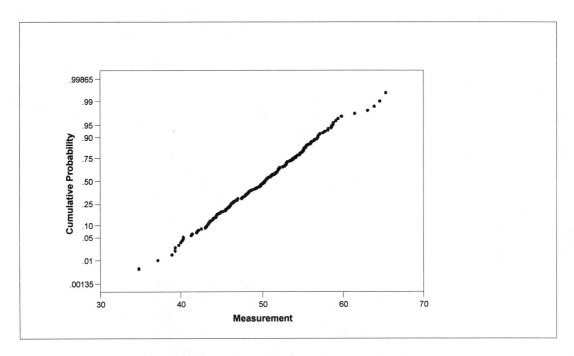

Figure 2.9 Probability Plot with Synthesized Normal Data ($n = 200$)

If the distribution is moderately skewed to the right, a sample's data will yield a probability plot that is slightly convex toward the upper side, similar to the one shown in Figure 2.10, which plots the data from the histogram in Figure 2.6. If the distribution is severely skewed to the right, a sample's data will yield a probability plot that is more severely curved, similar to the one shown in Figure 2.11, which plots the data from the histogram in Figure 2.7.

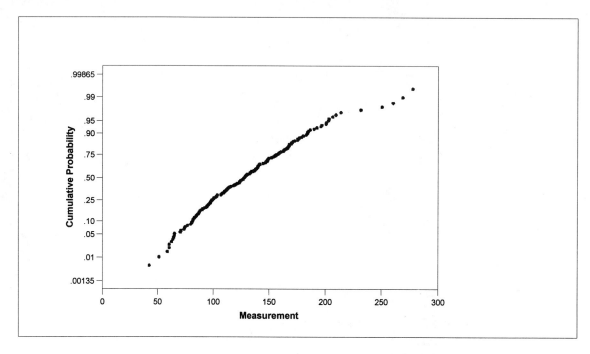

Figure 2.10 Probability Plot of Data That Is Moderately Skewed to the Right ($n = 200$)

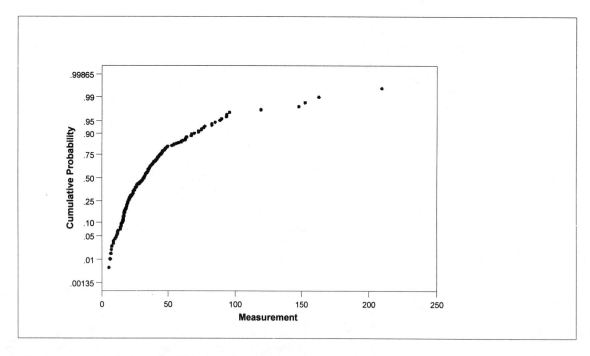

Figure 2.11 Probability Plot of Data That Is Severely Skewed to the Right ($n = 200$)

The probability plot is a better graphical test for near-normality than is a histogram. The shape of the histogram is highly dependent upon the size and number of cells. This is not a problem with the probability plot. How straight does the line have to be to consider it near-normal? As Shapiro [1990, p. 9] points out:

> **If the model is appropriate** then the plotted points will tend to fall on a straight line. **If it is not appropriate** the points will deviate from a straight line, generally in some systematic manner. The decision whether or not to reject the hypothesized model is subjective and two people looking at the same data might come to different conclusions, but with some experience a reasonably good assessment can be made.

As mentioned earlier, the probability plot is also an excellent tool for estimating the percentage of the population in the tails of the distribution and the process capability. This is true whether the data are near-normal or not. This aspect of the probability plot will be discussed in detail in Chapter 6.

Chapter Summary

Average:
- The average is a descriptive statistic—a measure of central tendency.

- The most common measure of central tendency is the mean or the "arithmetic" average, called here simply the average.

- The symbol for the mean is $\overline{X}$ or Xbar (read "X bar").

- Xbar is computed by adding the values and then dividing by the number of values, that is,

$$Xbar = \frac{\sum_{i=1}^{n} X_i}{n}$$

Range:
- The range is a descriptive statistic—a measure of variation that uses only the two extreme values (the highest and the lowest).

- The range is symbolized by R.

- Formula: R = Highest – Lowest.

Standard deviation:
- The standard deviation is a descriptive statistic—a measure of variation that uses all the data.

- Most statistical tests use standard deviation.

- The standard deviation of the sample is symbolized by s; the population standard deviation is symbolized by σ.

- Formula:

$$s = \sqrt{\dfrac{\sum\limits_{i=1}^{n}(X_i - Xbar)^2}{(n-1)}}$$

where n is the number of observations $X_1, X_2, \ldots, X_n$.

Normal distribution:
- A normal distribution appears to be bell-shaped.

- Almost all (99.73%) of normally distributed data are within three standard deviations of the average, so about one in 1,000 will fall in each tail beyond three standard deviations from the average.

- A near-normal distribution is one where departures from normality will have no detrimental effect on the control chart.

Histogram:
- The histogram graphs the frequency of each measurement.

- A histogram is used to get a "quick picture" of the shape of the distribution.

- A histogram often groups the data into cells or bins.

- The bins reflect the fact that the data are continuous, that is, where one cell ends, the next begins.

Probability plot:
- The probability plot graphs the relative cumulative frequency on a special scale. Near-normally distributed data will plot approximately as a straight line.

- The probability plot helps determine the general shape of the distribution.

- The probability plot facilitates the estimation of the percentage of the data in the tails of the distribution and the process capability.

<u>Problems</u>

1. Calculate the average (Xbar), range, and standard deviation for each of the following sets of data.
    a.   5, 8, 3, 14
    b.   4, 3, 3, 6, 8

2. Calculate the average (Xbar), range, and standard deviation for each of the following sets of data.
    a.   9, 11, 6
    b.   0.83, 0.79, 0.64, 1.23, 0.98

3. Make a histogram for the following sets of data.
    a.   2 3 5 3 4 5 7 2 5 5 1 2 7 2 3 2 2 1 4
    b.   250 100 75 75 70 155 65 60 75 70 88 90 115 90 75 75 120 125 85 80 105 95 80 85
         150 100 85 75 80 80 160 85 110 100 120 210 85 90 270

4. Make a histogram for the following sets of data.
    a.   110 120 100 90 105 130 75 95 105 103 130 60 60 85 125 120 70 60 60 80
         105 80 55 125 60 100 125 62 95 90 75 80 93 100 60 70 80 110 65 90
    b.   4 8 3 2 5 7 6 11 2 7 0 1 13 22 15 2 11 3 5 2 4 2 3 2 2 8 1 6 4 1 0 11 0 4 6 19 1 16
         0 2 20 1 7 4 2 5 6 2 1 10 3 17 3 10 30 9 5 1 3 25 8 26 34 7 5 2 4 21

5. Calculate the relative cumulative frequencies using the plotting convention $i/(n + 1)$ required to make a probability plot for each of the data sets in problem 3.

6. Calculate the relative cumulative frequencies using the plotting convention $i/(n + 1)$ required to make a probability plot for each of the data sets in problem 4.

Computer Supplement (Statit)

Preparing the Data

In Statit Express QC, a column is a "variable" and a row is a "case." All of the data for one variable must be in one column. For example, suppose it is desirable to use surgery times for given dates in time order. The first variable may be the date, the second column the surgery times, as shown below. (This is actual data from one hospital.) Note that the variable names are Date and Surgery_minutes, each one word.

EXAMPLE 1

This is an example of surgery minutes. You will key this data in later in Example 2.

	Date	Surgery_minutes
1	1-4-01	110
2	1-5-01	120
3	1-9-01	100
4	1-23-01	90
5	1-25-01	105
6	1-31-01	130
7	2-14-01	75
8	3-12-01	95
9	5-4-01	105
10	5-5-01	103

Note that all of the surgery times are kept in one column. Other complications, such as adding a column for surgeon identification, will be addressed in later examples. It is recommended that each case (row) have a unique identifier, such as patient id, row number, and so on. In this case, note that each date is different, providing a unique identifier for each row. □

Entering Data

To do statistical analysis of data, it is obvious that data are needed. There are three primary methods to input data into Statit Express QC:

1. Keying the data directly into Statit Express QC.
2. Opening a data set previously saved in Statit Express QC.
3. Importing the data from a spreadsheet such as Excel.

These methods are explained in detail below.

EXAMPLE 2

To enter the data from Example 1 directly into Statit Express QC, click on
        Window > Data
to make the Data Window active. Then choose
                Edit > New Variable . . .
                        Variable Name: type in *Date*
                        Data Type: √ Date
                        OK
                Edit > New Variable . . .
                        Variable Name: type in *Surgery_minutes*
                        Data Type: √ Numeric
                        OK

Date Format: Click on the first cell, then key in the data as you would in a spreadsheet. Enter the entire data set from Example 1. Note that as you key in the date 1-4-01, it appears as 04-Jan-01 (or similar format). When all of the data are entered, change the print format of the date by doing the following:
                Right-click on the variable name (column heading) Date
                Format
                Date
                        select the format you desire
                OK ☐

EXAMPLE 3

Saving the Data:
        File > Save Data File . . .
                Save in: (click on ▼, then on 3 1/2 Floppy (A:) if desired)
                File name: type in *sem1* (or sem1.wrk—Statit will add the extension .wrk if you don't)
                Save

Note: You can also skip right to File name and type in *A:\sem1.wrk*. ☐

EXAMPLE 4

Opening Data Previously Saved:
To reopen sem1.wrk (for practice in opening files), choose
        File > Open Data File . . .
                Look in: (click on ▼, then on 3 1/2 Floppy (A:) if desired)
                click on the file you wish to open and it will appear in File name:

Open

Note: You can also skip right to File name and type in *A:\sem1.wrk*. □

Importing Data from a Spreadsheet

To import data from a spreadsheet such as Excel, you must first "clean up" the data in the spreadsheet. You can have at most one row for column headings above the data. The column headings, if any, will become the variable names.

Then select

> File > Open Data File . . .
> > Look in: click on drive and directories needed
> > Files of type: click on Excel Files (*.xls) or such for your program file
> > click on the file name needed
>
> Open
>
> (Alternate approach: key in the complete file name such as C:\Excel\dataset.xls, then click on Open.)

EXAMPLE 5

Calculating Average and Standard Deviation:
Statistics > Descriptive . . .
> Variable: (click on ▶, Surgery_minutes, Done)
> Output Format: √ at Table
> Statistics: remove √ at some of the statistics listed,
> > leave mean, standard deviation and some other selected statistics checked
>
> OK □

EXAMPLE 6

Histogram:
To make a histogram on the surgery times from the data set sem1.wrk, choose
> Graphs > Histogram . . .
> > Data Variable: (click on ▶, Surgery_minutes, Done)
> > Chart Title: type in *Lap Chole Surgery Times*
> > Sub title: type in *Jan - May*
>
> X-Axis title: type in *Surgery Times in Minutes*
> Y-Axis title: (Leave blank. It will default to "Frequency.")
> OK

Note: Statit's intervals include right endpoints. □

EXAMPLE 7

To make a histogram on the surgery times from the data set sem1.wrk with 6 bins, choose
    Graphs > Histogram . . .
        Data Variable: (click on ▶, Surgery_minutes, Done)
        Number of bins: type in *6*
        Chart Title: type in *Lap Chole Surgery Times*
        Sub title: type in *Jan - May*
    X-Axis title: type in *Surgery Times in Minutes*
    Y-Axis title: (Leave blank. It will default to "Frequency.")
    OK ☐

EXAMPLE 8

Probability Plot:
With data set: sem1.wrk

    Graphs > Probability . . .
        Variable: (click on ▶, Surgery_minutes, Done)
        delete √ at Plot a theoretical quantiles line
        delete √ at Perform test for normality
        Axis endpoints:
            √ at .00135 - .99865
        OK ☐

Computer Supplement (Minitab)

Preparing the Data

In the student edition of Minitab, a column is a "variable" and a row is a "case." All of the data for one variable must be in one column. For example, suppose it is desirable to use surgery times for given dates in time order. The first variable may be the date, the second column the surgery times, as shown below. (This is actual data from one hospital.)

EXAMPLE 1

This is an example of surgery minutes. You will key this data in later in Example 2.

	Date	Surgery_minutes
1	1-4-01	110
2	1-5-01	120
3	1-9-01	100
4	1-23-01	90
5	1-25-01	105
6	1-31-01	130
7	2-14-01	75
8	3-12-01	95
9	5-4-01	105
10	5-5-01	103

Note that all of the surgery times are kept in one column. Other complications, such as adding a column for surgeon identification, will be addressed in later examples. It is recommended that each case (row) have a unique identifier, such as patient id, row number, and so on. In this case, note that each date is different, providing a unique identifier for each row. □

Entering Data

To do statistical analysis of data, it is obvious that data are needed. There are three primary methods to input data into Minitab:

1. Keying the data directly into Minitab.
2. Opening a data set previously saved in Minitab.
3. Importing the data from a spreadsheet such as Excel.

40

These methods are explained in detail below.

EXAMPLE 2

To enter the data from Example 1 directly into Minitab, click on the spreadsheet, then type in the data with the column headings on the top line. □

EXAMPLE 3

Saving Data:
       File > Save Current Worksheet As . . .
              Save in: (click on ▼, then on 3 1/2 Floppy (A:) if desired)
              File name: type in *sem1*
              Save

(Minitab puts on the extension .mtw.) □

EXAMPLE 4

Opening Data Previously Saved:
To reopen sem1. (for practice in opening files), choose
       File > Open Worksheet . . .
              Look in: (click on ▼, then on 3 1/2 Floppy (A:) if desired)
              click on the file you wish to open and it will appear in File name:
              Open
(Here: Reopen? (click on Yes)) □

Importing Data from a Spreadsheet

To import data from a spreadsheet such as Excel, you must first "clean up" the data in the spreadsheet. You can have at most one row for column headings above the data. The column headings, if any, will become the variable names.

Then select
       File > Open Worksheet . . .
              Look in: click on drive, directories needed
              Files of type: click on ▼, Excel (*.xls)
              Click on file name needed

       Open

EXAMPLE 5

Calculating Average and Standard Deviation:
Stat > Basic Statistics > Display Descriptive Statistics . . .
   Variable: (click on C2 Surgery_minutes in left column, Select)
   OK □

EXAMPLE 6

Histogram:
To make a histogram on the surgery times from the data set sem1, choose
   Graph > Histogram . . .
      Graph Variables: (click on C2 Surgery_minutes, Select)
      Annotation (click on▼, Title . . .)
         Title: type in:
            1. *Lap Chole Surgery Times*
            2. *Jan – May*
        OK
      Options . . .
         Type of Histogram:
           √ at Frequency
         Type of interval:
           √ at Cutpoint
         Definition of Intervals:
           √ at Midpoint/*Cutpoint* positions
             then specify all in order from smallest to largest or
             shorthand (e.g., 70:140/10 means intervals from 70 to 140 by
             10)
        OK
     OK

Note: Minitab's intervals include left endpoints. □

EXAMPLE 7

To make a histogram on the surgery times from the data set sem1 by 5s, choose
   Graph > Histogram . . .
      Graph Variables: (click on C2 Surgery_minutes, Select)
      Annotation (click on ▼, Title . . .)
         Title: type in:
            1. *Lap Chole Surgery Times*
            2. *Jan - May*
        OK
      Options . . .

Type of Histogram:
                √ at Frequency
            Type of interval:
                √ at Cutpoint
            Definition of Intervals:
                √ at Midpoint/Cutpoint positions
                    type in *70:140/5*
    OK ☐

EXAMPLE 8

Probability Plot:
With data set: sem1

        Graph > Probability Plot . . .
            Variables: (click on C2 Surgery_minutes, Select)
            Distribution: Normal
            Options . . .
                delete √ at Include confidence intervals in plot
                OK
            OK

Note: Minitab uses the plotting convention $(i - 3/8)/(n + 1/4)$. ☐

Chapter 3 The Run Chart for Time-Ordered Variables Data

The Simplest SPC Chart: The Run Chart

The basic *run chart* is a plot of the individual variables data observations in time order. It is the simplest chart for SPC, and it is also one of the most useful. Traditional statistics refers to such a run chart as a *time-series plot*. The run chart is very useful for detecting trends or patterns in the time-ordered data. Unlike the control chart, which needs a near-normal distribution of the variables data, the run chart is more forgiving of departures from normality. In fact, reference is made at the end of this chapter to the use of run charts for applications other than the individual observations in variables data. The run chart will usually display its centerline (CL), Xbar, which is the average of the observations and is the estimate for the overall level of the process.

EXAMPLE 3.1

In order to schedule laparoscopic cholecystectomy (lap chole) surgeries, a study was made of the surgery times in minutes. Table 3.1 gives surgery times for 40 consecutive lap choles performed by a new surgeon. These are displayed on the run chart in Figure 3.1. The Xbar of 90.1 minutes is denoted in the chart as the centerline.

The run chart in Figure 3.1 shows no indications of lack of randomness; it could just as well have come from a table of random normal numbers as from real process data. However, since the run chart has no statistically derived control limits, it lacks the power of the control chart to detect evidence of nonrandomness. □

Table 3.1 Consecutive Laparoscopic Cholecystectomy (Lap Chole) Surgery Times in Minutes

Surgery Number	1	2	3	4	5	6	7	8	9	10
Surgery Time	110	120	100	90	105	130	75	95	105	103

Surgery Number	11	12	13	14	15	16	17	18	19	20
Surgery Time	130	60	60	85	125	120	70	60	60	80

Surgery Number	21	22	23	24	25	26	27	28	29	30
Surgery Time	105	80	55	125	60	100	125	62	95	90

Surgery Number	31	32	33	34	35	36	37	38	39	40
Surgery Time	75	80	93	100	60	70	80	110	65	90

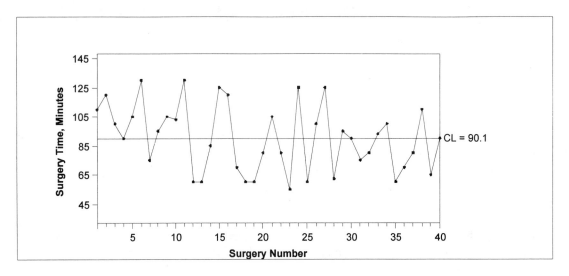

Figure 3.1 Run Chart on 40 Consecutive Lap Chole Surgery Times

Time-Ordered Analysis and Single-Stream Processes

For any useful analysis of time-ordered data (whether from a run chart, a control chart, or another method), the data must be homogeneous—from a *single-stream process* (e.g., from the same hospital or from the same shift within a hospital). Unfortunately, sometimes it may be difficult to sort out a single-stream process. For example, different workers within a shift may be acting as different processes. Best efforts must be made to sort out exactly what a single-stream process is. Deming [1982] points out that to improve the quality of the river, one should not study the river but must go upstream and study the individual streams that feed the river.

To illustrate the importance of the single-stream process, consider the following scenario. A healthcare organization owns two hospitals, hospital A and hospital B, and wants to track the mortality rate by month for a given procedure. If mortality is trending upward in hospital A and downward in hospital B, the combined rate for the two hospitals may be steady over time, or trending upward, or trending downward. The combined rate for the two streams has no meaning unless the two streams are alike. Each hospital must be analyzed separately. If the two data streams are found to be acting similarly on the same level, they may be combined, or *pooled*. See Appendix 2 for further discussion on the perils of pooling.

Historical Analyses and Ongoing Process Monitoring: No Standard Given and Standard Given

The time-ordered data in Table 3.1 that were referred to above are historical (retrospective) data; the complete data set was collected in the past. These data were analyzed with a run chart in Example 3.1 using Xbar, the average of that data set, as the centerline. Since the analysis was done without reference to any other data or to any other "outside standards" (e.g., standards from a governmental or accreditation organization), it is said that there was *no standard given*.

However, a *standard value* might be used for the centerline instead of calculating it from the data set. When data are analyzed using a standard value, the analysis is said to be with *standard given* [ANSI, 1975]. For instance, the centerline may come from data collected earlier. This is often done for *ongoing process monitoring* (prospective as opposed to retrospective analysis). For ongoing process monitoring, the centerline is projected into the future so that, as new data are collected, they are plotted in real time on the run chart. This way, a lack of randomness is more likely to be discovered promptly so that a search may be made immediately for the special cause of variation. This greatly aids process improvement.

Standard values can also come from data collected from another application or from an outside standard. This is generally done for comparison—to see how the process currently being studied compares to some standard value.

EXAMPLE 3.2

The surgery times in Table 3.1 are from a new surgeon. The mean surgery time for all existing surgeons is 70 minutes, which is to be used as a standard value. The surgery times for the new surgeon are plotted on the run chart in Figure 3.2 using the 70 minutes as the "standard given." From this chart it appears that the new surgeon's surgery times tend to be higher than the others. Formal criteria to determine that this is the case will be discussed in the next section. Longer times must be scheduled for the new surgeon. □

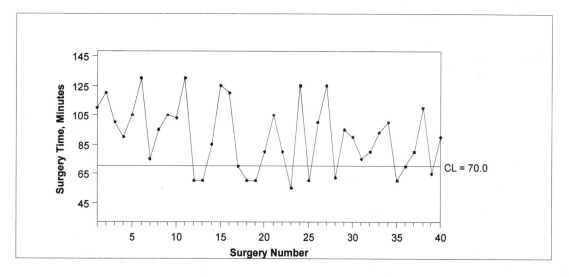

Figure 3.2 Run Chart on Lap Chole Surgery Times, Standard Given

Use of a Run Chart

A run chart is used to monitor a process. Consequently, it looks for patterns, trends, obviously excessive points, and so on, that suggest nonrandom influence. Because it does not have statistically derived limits, it is sometimes difficult to determine exactly what an obviously excessive point is. The control limits will calculate statistically derived limits that will help discern what is meant by "excessive."

When deciding if the run chart suggests nonrandom influence, two types of errors may be made. Sometimes the data truly come from a random process, but by pure chance a nonrandom pattern occurs. The process is mistakenly declared to be nonrandom, so a *type I error* has been made. The probability of such an error is called the *type I risk, false alarm risk,* or *α-risk* (alpha risk). On the other hand, the data may come from a process that really has nonrandom influence, but no evidence of this shows up on the chart. The process is mistakenly declared to be random, so a *type II error* has been made.

Some formal tests for patterns that suggest nonrandom influence exist. They include the following:

1. A run of "too many" points on the same side of the centerline. How many points are needed to be "too many" depends on the number of points plotted and the α-risk you are willing to accept. For an α-risk of around 0.05, runs of consecutive points on one side of the centerline as long or longer than the values in Table 3.2 suggest nonrandom influence in *n* points [adapted from Duncan, 1986]. For a "typical" chart of 24 to 30 points, this leads to the often quoted tests of eight points [Western Electric/AT&T, 1956, p. 26] or nine points [Nelson 1984, p. 238] on the same side of the centerline. (See Figure 3.3.)

Note: This and similar tests were originally designed for the centerline being the *median*, that is, the middle observation after the observations have been arranged from smallest to largest. Now Xbar is more favored. If the distribution is reasonably symmetric, the median approximately equals Xbar.

Table 3.2 Length of Runs on One Side of the Centerline Suggesting Nonrandom Influence

n	Run Length
10 or fewer	6
15 or fewer	7
20 or fewer	8
30 or fewer	9
40 or fewer	10
50 or fewer	11

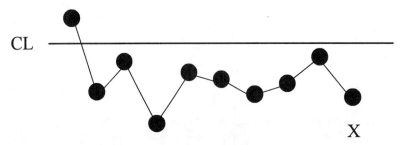

Figure 3.3 Nine Consecutive Points below the Centerline

2. A run of points steadily increasing (or decreasing) that is "too long." For an α-risk of around 0.05, runs of consecutive points increasing (or decreasing) as long or longer than the values in Table 3.3 suggest nonrandom influence in n points [adapted from Duncan, 1986]. For a "typical" chart of 24 to 30 points, this leads to the often quoted test of six points in a row (including endpoints) steadily increasing (or decreasing) [Nelson, 1984]. (See Figure 3.4.) A rule sometimes used and recommended here is that points that are equal to the previous reading are ignored and not counted in either the number of points or the increases (decreases).

Table 3.3 Length of Runs Steadily Increasing (Decreasing), Giving Evidence of Nonrandomness

n	Run Length
6 – 8	5
9 – 30	6
31 – 150	7
150 – 1000	8

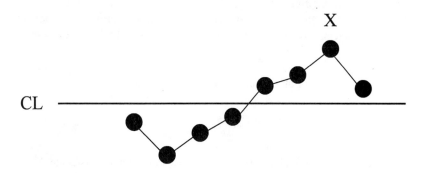

Figure 3.4 Six Consecutive Points Constantly Increasing

3. "Too few" points on one side of the centerline. Table 3.4 may be used to determine how few points must be on one side of the given centerline out of n points to suggest nonrandom influence with an α-risk less than 0.09.

Table 3.4 Number of Points on One Side of a Given Centerline Giving Evidence of Nonrandom Variation

n	Total Count on One Side of the Centerline
8 or more	1 or fewer
12 or more	2 or fewer
14 or more	3 or fewer
16 or more	4 or fewer
19 or more	5 or fewer
21 or more	6 or fewer
24 or more	7 or fewer

The Interrupted Run Chart

A series of data in time order will often reflect interruptions, either to the process or to the continuous sampling from the process. It is important that the run chart emphasize these interruptions. This is done by connecting the points in time order *except* where there is an interruption. The term *interrupted run chart* is used here to emphasize the advantages of selectively interrupting (i.e., disconnecting) the lines between the plotted points.

The interrupted run chart is used to discover nonrandomness in Case Studies 4.1 and 5.1.

Run Charts with Other Types of Data

The run charts considered here have been for individual observations with variables data. This is the primary application, but the run chart may sometimes also be used to advantage with either variables or attribute data to plot any of the subgroup statistics in time order.

Chapter Summary

- A run chart is a time-ordered plot, usually of individual observations.

- For the usual case of no standard given, the run chart centerline, Xbar, is the average of the observations.

- The run chart is most effective if made on a time-ordered set of homogeneous data from a single-stream process.

- A standard value could come from one of three sources:

    1. Data collected earlier on this process (to detect special-cause variation to aid process improvement)
    2. Data collected from another application (for comparison)
    3. An outside standard (for comparison)

- For ongoing process monitoring, the centerline is projected into the future so that, as new data are collected, they are plotted in real time on the run chart.

- When testing for process instability, type I errors (falsely indicating instability) must be balanced against type II errors (failure to detect instability).

- Three formal tests for evidence of nonrandomness are

    1. A run of "too many" points on the same side of the centerline
    2. "Too many" consecutive points (including endpoints) steadily increasing or decreasing
    3. "Too few" points on one side of the centerline

- When there are interruptions to the process or to the data acquisition, an *interrupted run chart* should be used, selectively disconnecting the lines between the plotted points.

- The run chart may be used to plot any subgroup statistics in time order with either variables or attribute data.

Problems

1. Make a run chart using only the first 8 values in Table 3.1 and a standard value of 85 for the centerline. Is there evidence of special-cause variation? Is there sufficient data for this chart?

2. Make a run chart using only the first 8 values in Table 3.1 and a standard value of 80 for the centerline. Is there evidence of special-cause variation? Is there sufficient data for this chart?

3. Make a run chart on the following 60 time-ordered surgery times (in minutes) for lap choles by surgeon A. What do you see from the data?
65 75 75 60 165 75 85 85 80 95 65 65 85 68 190 120 105 115 58 70 80 75 65 75 90 75 65 50 80 75 80 115 75 95 130 125 105 70 72 95 90 120 75 90 85 90 80 115 80 65 70 185 75 90 120 130 65 65 115 65

4. Make a run chart on the following time-ordered surgery times (in minutes) for lap choles by a particular surgeon. What do you see from the data?
250 100 75 75 70 155 65 60 75 70 88 90 115 90 75 75 120 125 85 80 105 95 80 85 150 100 85 75 80 80 160 85 110 100 120 210 85 90 270

5. Make a run chart on the following time-ordered surgery times (in minutes) for lap choles by a particular surgeon. What do you see from the data?
    120 110 110 95 100 135 75 95 105 110 130 60 65 85 125 120 75 70 60 80
    105 85 75 115 60 100 120 60 95 90 70 80 95 100 65 70 80 110 65 100

6. Make a run chart on the following number of days between surgical site infections for a given surgery. What do you see from the data?
    4 8 3 2 5 7 6 11 2 7 0 1 13 22 15 2 11 3 5 2 4 2 3 2 2 8 1 6 4 1 0 11 0 4 6 19 1 16
    0 2 20 1 7 4 2 5 6 2 1 10 3 17 3 10 30 9 5 1 3 25 8 26 34 7 5 2 4 21

7. Make an interrupted run chart from the following data. (A ? means an interruption.) What do you conclude?
    120 110 110 95 100 ? 135 75 95 105 110 130 ? 60 65 85 125 120 75 70 60 80

8. Make an interrupted run chart from the following data. (A ? means an interruption.) What do you conclude?
    105 85 75 115 60 ? 100 120 ? 60 95 90 70 80 95 100 ? 65 70 80 110 65 100

Computer Supplement (Statit)

EXAMPLE 1

Run Chart:
To make a run chart, with data set: sem1.wrk

> [I] button (stands for Individual)
>> Data Variable: (click on ►, Surgery_minutes, Done)
>> X Labels variable: (click on ►, Date, Done)
>> Individual Chart Control limits:
>>> at Upper: (click on ▼, Off)
>>> at Lower: (click on ▼, Off)
>> X-Axis label frequency: (type in *1*)
>> Chart title: (type in *Surgery Times*)
>> Sub title: (type in *minutes*)
>> OK ☐

Interrupted Run Chart

Statit uses a ? (the missing value symbol) to denote an interruption in the process.

EXAMPLE 2

With data set: sem3.wrk:

	Date	Surgery_minutes
1	1-4-01	110
2	1-5-01	120
3	1-9-01	100
4	1-23-01	90
5	1-25-01	105
6	1-31-01	130
7	2-1-01	?
8	2-14-01	75
9	3-12-01	95
10	5-4-01	105
11	5-5-01	103

[I] button

        Data Variable: (click on ▶, Surgery_minutes, Done)

        X Labels variable: (click on ▶, Date, Done)

        Individual Chart Control limits:

                at Upper: (click on ▼, Off)

                at Lower: (click on ▼, Off)

        X-Axis label frequency: (type in *1*)

        Chart title: (type in *Surgery Times*)

        Sub title: (type in *minutes*)

        OK □

Computer Supplement (Minitab)

EXAMPLE 1

Run Chart:
To make a run chart, with data set: sem1
       Graph > Time Series Plot . . .
             Graph Variable: (click on C2 Surgery_minutes, Select)
       OK □

<u>Interrupted Run Chart</u>

Minitab uses a * (the missing value symbol) to denote an interruption in the process. However, Minitab does not interrupt the plot for an interruption on a run chart.

Chapter 4 Control Chart Theory and the I Chart for Time-Ordered Data

Theory Applicable to All Control Charts

In Chapter 1 it was seen that there are two types of variation: *common-cause variation* that is common to the whole process and *special-cause variation* that is a signal of an unusual event. It was further noted that a control chart is a statistical tool used primarily in the quality area to discriminate between special-cause variation and common-cause variation. For the usual case of "no standard given," this is done with control limits calculated from the data to estimate the extent of common-cause variation. Traditionally, these control limits have been three standard deviation (3-sigma or 3σ) limits, which would be expected to include 99.73% of the plotted points if the data were normally distributed. But even though real data cannot be perfectly normally distributed, 3-sigma limits are still used because they "have been found to give good practical results" [Duncan, 1986, p. 424]. When the calculated values from the samples fall within the limits on the control chart, it is assumed that only the common-cause variation is occurring, and the process is said to be in a "state of statistical control," "in control," or "statistically uniform."

Recall from Chapter 1 that Shewhart [1931, p. 6] defined control by saying:

> … a phenomenon will be said to be controlled when, through the use of past experience, we can predict, at least within limits, how the phenomenon may be expected to vary in the future. Here it is understood that prediction within limits means that we can state, at least approximately, the probability that the observed phenomenon will fall within the given limits.

The British Standard BS 600 written by Pearson [1935, p. 38] has used the term *statistically uniform* to describe this state of stability

> to avoid the confusion which has found to occur when using the words "statistical control" between the concept of statistical uniformity and the pragmatic interpretation of the phrase as an act, for example, of a process manager.

When a plotted point falls outside the limits on the control chart, it is a signal that a special cause of variation is probably present and needs to be found. Deming [1982, p. 112] warns:

> A statistical chart detects the existence of a cause of variation that lies outside the system. It does not find the cause.

Once the special cause is found, action must be taken to eliminate that special cause of variation if it is detrimental to the process, to make it standard procedure if it yields a desired effect, or at the very least, to take it into account in the analysis. As Juran [1945, pp. 120 – 121] points out:

The control chart technique is admirably suited to executive review. This executive review carries with it the responsibility to see to it that corrective action is taken where such action is indicated. In the absence of corrective action, the control chart technique deteriorates into a sterile paper work procedure!

Furthermore, the control chart identifies when only the natural common-cause variation of the process is present, in which case action may result in more variability. If adjustments are made when only common-cause variation is present, these adjustments move the whole process distribution back and forth, resulting in increased variability. In other words, the control chart tells when to take action on the process and when to leave it alone.

If the control chart shows no evidence of special-cause variation, it does not mean that none is present. It may be that not enough data were collected, the data were not collected properly, the data were not subgrouped properly, or chance in sampling yielded only the in-control data. As Grant and Leavenworth [1996, p. 46] put it:

> When we say, "This process is in control," the statement really means, "For practical purposes, it pays to act as if no assignable causes of variation were present."

Alternately, if the control chart exhibits evidence of special-cause variation, it does not mean for certain that a special cause of variation exists. Recall from Chapter 2 that for a normal distribution, 99.73% of the data will fall within three standard deviations from Xbar, implying that about 3 points in 1,000 will fall beyond three standard deviations from Xbar. Therefore, the process may falsely exhibit evidence of special-cause variation when none is present. As noted in Chapter 3, this is referred to as a *false alarm* (or *type I error*). The probability of such a false signal happening is called a *false-alarm risk* or *alpha (α-) risk*.

As with the run chart, a control chart for time-ordered data should only be made with data that behave as a "single-stream" process; that is, the data were all gathered from the same shift at the same hospital, or from the same DRG group, or from the same procedure by the same surgeon. More than one surgeon may be kept on the same chart only after a study finds no significant statistical difference between these surgeons.

Another important method of data portrayal is to subgroup the data "cross-sectionally" or in "rational subgroups" for the comparison of several "streams" or strata. For example, five hospitals (subgroups) might be compared using 24 observations each. Both variables data and attribute data are analyzed in time order and by using rational subgroups.

The following suggestions are made for the minimum number of plotted points to be used for time-ordered data. These are intended for all of the control charts illustrated in this book (I, Xbar, s, c, u, and p charts):

1. Ten plotted points may be sufficient for process improvement if their results suggest nonrandom influence, since discovering such influence is the first goal. If the control chart results do not suggest nonrandom influence, they may still be useful for process improvement to provide *trial control limits* pending the accumulation of more data.

2.   Usually, it may not be safely inferred that a state of control exists unless 24 consecutive plotted points fall within the control limits.

Variables Control Charts

Variables control charts are control charts that use variables data (rather than attribute data). Variables data are measurement data and include measurements and percentages derived from measurements, such as the percentage of solids in a solution. These quantities exist on a continuous scale or on a scale that theoretically could be divided into an infinite number of increments. However, the actual measurement scale can only be divided into a finite number of increments; variables data are always rounded to some increment.

Some healthcare data may be satisfactorily near-normal. Examples are blood pressure (systolic and diastolic in mm of Hg), and cost (which might be rounded to the nearest thousand dollars, to the nearest dollar, etc.). However, time intervals may be too severely skewed to the right to qualify as "near-normal."

The control limits for variables data are derived assuming that the data are normally distributed. When a point falls above the upper control limit on a variables control chart, there is evidence of special-cause variation if the distribution is near-normal, but if the distribution is badly skewed to the right, it is impossible to determine whether the outage is due to the skewed distribution or to special-cause variation.

As noted in Chapter 2, the probability plot is used to determine whether the data may be classified as near-normal. The probability plot may often be advantageously supplemented with a histogram to check this near-normal assumption. If the data are not satisfactorily near-normal, an appropriate transformation will be required before making a control chart, as will be discussed in Chapter 9.

The control limits for variables control charts are calculated from a "within-subgroup estimate" of the standard deviation. This is fundamentally different from the practice in classical statistics where the "overall" value of s (discussed in Chapter 2) is used, calculated from all of the data. The "within-subgroup estimate" of sigma estimates the standard deviation only from observations within a subgroup—which are expected to be very much alike. It follows that the within-subgroup estimate of sigma tends to be smaller than the overall estimate of s, yielding *control limits* that tend to be tight (narrow), and hence, useful in the detection of special-cause variation.

"T-Sigma Limits" versus 3-Sigma Limits

Conventional 3-sigma control limits are usually used for data sets with about 25 subgroups. For such applications the 3-sigma limits provide a good balance between providing sufficient power for the detection of special-cause variation and maintaining a satisfactory total α-risk of getting one or more points outside the control limits by random chance.

However, 3-sigma limits do not maintain a satisfactory α-risk when the number of subgroups differs greatly from 25. About one in fifteen charts with 25 points each would wrongly display evidence of lack of statistical control.

When control charts are maintained for ongoing process improvement over an extended period of time, the number of subgroups may become quite large. Given enough subgroups, a perfectly stable process will have points outside the 3σ limits simply by random chance. This problem has been acknowledged in the American National Standards Institute (ANSI) Standard Z1.3 [1958, 1975, p. 18]. It states that

> it is usually not safe to conclude that a state of control exists unless the plotted points for at least 25 successive subgroups fall within the 3-sigma control limits. In addition, if not more than 1 out of 35 successive points, or not more than 2 out of 100, fall outside the 3-sigma control limits, a state of control may ordinarily be assumed to exist.

Although this is a very useful guideline, no mathematical basis is given. Also, no suggestions are given for the evaluation of more than 100 subgroups, nor is the problem of cross-sectional (as opposed to time-ordered or longitudinal) data considered, where there may be as few as two subgroups (e.g., comparing results from two shifts), which will be discussed in Chapter 5.

A method for maintaining a satisfactory α-risk is to use control limits other than 3σ limits, which may vary with the number of subgroups and possibly with the subgroup size. The objective is to maintain the highest possible power in detecting special-cause variation without an "excessive" alpha risk.

A simple approach is to use 2-sigma limits for 12 or fewer subgroups [Hart and Hart, 1989]. *Analysis of means,* developed by Ellis Ott [Ott, 1975; Schilling, 1973], is a more sophisticated method of selecting the control limits to be used for a given alpha risk. However, since analysis of means depends on both subgroup size and the number of subgroups, it requires extensive tables and is seldom used in practice. Studies done by Hart and Hart [1994, March 1996, April 1996, 1997] have simplified the methodology by the use of "T-sigma" limits, which keep the total α-risk to less than 0.09. The results are given in Table 4.1.

Table 4.1 Recommended Values of T for T-Sigma Limits*

# of plotted points	T
2	1.5
3 – 4	2.0
5 – 9	2.5
10 – 34	3.0
35 – 199	3.5
200 – 1500	4.0

* The tabular values of T may be used for the usual case of "no standard given" with all attribute and variables charts. For "standard given," use T = 2 for 1 or 2 subgroups. For 3 or more subgroups the tabular values may be used.

Note that 3-sigma limits are typically used for ongoing process improvement. For process improvement on critical processes, 2-sigma limits are sometimes used, feeling that the increased α-risk is worth the trouble because of a better chance of finding problems early. For more regarding ongoing process improvement, see the section on "standard given" later in this chapter.

For ease of calculation, examples in this book where detailed hand calculations are shown use the traditional 3-sigma limits, regardless of the number of subgroups. T-sigma upper and lower control limits (T-sigma UCL and T-sigma LCL) may be calculated by hand by first calculating the 3-sigma limits (UCL and LCL) and then by adding and subtracting the value of T-sigma to the chart centerline (CL), where T-sigma = (T)(UCL - CL)/3. Unless labeled otherwise, the control limits are assumed to be 3-sigma limits. Case studies, where the analysis is done by computer, will use T-sigma limits.

Use of the I Chart

The I chart, also called an Individual Chart or an X chart (where X stands for a data value), is simply a run chart with control limits added. It is simple to construct and interpret, so healthcare personnel responsible for the chart quickly become comfortable with it and assume ownership of it. Like the run chart, the I chart displays the values of the process observations in time order. Also, as with the run chart, the average of the observations, Xbar, is the estimate for the overall level of the process and is used for the chart centerline. A long run of time-ordered data without interruptions is usually used for the I chart. (A modification of this chart will be introduced later in this chapter for use when there are interruptions in the process or in the data collection, since control limits will often be inflated if calculated across a process interruption.)

When data are analyzed in time order, the objective is to determine stability for the prediction of future performance. To calculate a within-subgroup estimate of standard deviation, the moving range (MR) is employed. An MR value is the difference between the current observation and the previous observation, high minus low. In expectation, these consecutive observations will be very much alike leading to small MR values. The average of the MR values, called *MRbar* (or $\overline{MR}$) is then used to calculate the within-subgroup estimate of process standard deviation. Quoting Lloyd Nelson [1982, p. 172]:

> The question frequently arises as to why the moving range of two is better than the standard deviation based on the whole preliminary sample, especially since this can be obtained so easily along with the mean on many pocket calculators. The reason is that the moving range of two minimizes inflationary effects on the variability which are caused by trends and oscillations that may be present. It measures variations from point to point irrespective of their average level.

The laparoscopic cholecystectomy (lap chole) surgery times in Table 3.1 are repeated in Table 4.2 with the MR values added. There is never an MR value associated with the first X value, so it is not applicable (NA). The first X value (surgery time) was 110 minutes and the second was 120. The MR value associated with the second X value is 120 - 110 = 10. The second X value was 120 and the third was 100, so the MR value associated with the third X value is 120 - 100 = 20. The number of MR values is one less than the number of points plotted on the I chart.

Table 4.2 Consecutive Laparoscopic Cholecystectomy (Lap Chole) Surgery Times in Minutes, with Moving Ranges

Surgery Number	1	2	3	4	5	6	7	8	9	10
Surgery Time	110	120	100	90	105	130	75	95	105	103
Moving Range	NA	10	20	10	15	25	55	20	10	2

Surgery Number	11	12	13	14	15	16	17	18	19	20
Surgery Time	130	60	60	85	125	120	70	60	60	80
Moving Range	27	70	0	25	40	5	50	10	0	20

Surgery Number	21	22	23	24	25	26	27	28	29	30
Surgery Time	105	80	55	125	60	100	125	62	95	90
Moving Range	25	25	25	70	65	40	25	63	33	5

Surgery Number	31	32	33	34	35	36	37	38	39	40
Surgery Time	75	80	93	100	60	70	80	110	65	90
Moving Range	15	5	13	7	40	10	10	30	45	25

The centerline of the I chart, as with the run chart, is Xbar. The I chart control limits, which show the limits of common-cause variation, are calculated as follows:

Upper Control Limit on X (i.e., on I):
$$UCL(X) = Xbar + (2.66)(MRbar)$$

Lower Control Limit on X:
$$LCL(X) = Xbar - (2.66)(MRbar)$$

The derivation of the formula may be found in Appendix 3. Recall that the derivation of the formulas assumes a normal distribution.

An MR control chart may be made on the MR values. However, it is suggested in this book that the MR chart *not* be made. Lloyd Nelson [1982, p. 173] suggests:

Advice varies on this point, but I recommend that the moving ranges not be plotted. Some delicacy of interpretation would be required because the moving ranges are correlated. Furthermore, the chart of the individual observations actually contains all the information available.

The 3-sigma control limits of the I chart have also been called the *natural limits* of the individuals. In Chapter 6 it will be noted that these provide a preliminary estimate of the *process capability limits* if the data are normally distributed.

The I Chart for Process Improvement

The most important use of the I chart is as a road map for process improvement, where the objective is to seek results that suggest nonrandom influence. The I chart for process improvement may have only ten or fewer points, but for ongoing process improvement it may have hundreds of points. The I chart will generally be made initially with no standard given. Even if a small number of points are used, a point outside the control limits would suggest the existence of nonrandom influence providing guidance for corrective action.

EXAMPLE 4.1

An initial process improvement effort is to be made at the point in time when only the first ten surgery times from Table 4.2 are available. The surgery times, recorded in minutes, are from consecutive laparoscopic cholecystectomies.

Before deciding to make any variables control chart, it is necessary to determine whether the data may be accepted as near-normal. The histogram and probability plots for the ten observations were shown in Figures 2.3 and 2.8. The probability plot indicates that a straight line would be a good fit; there is nothing in the histogram to the contrary, so the data are accepted as near-normal, and the next step—the making of the I chart—may properly be taken.

To calculate the I chart control limits, the following steps should be taken:

Step 1. Collect the individual measurements. The ten measurements are
    110   120   100   90   105   130   75   95   105   103

Step 2. Calculate the MR values.

X	110	120	100	90	105	130	75	95	105	103
MR		10	20	10	15	25	55	20	10	2

In this example, there are 10 points plotted on the I chart and 9 MR values.

Step 3. Find Xbar (the centerline for the I chart) and MRbar.

Xbar $= 1033/10 = 103.3$

MRbar $= 167/9 = 18.556$

Step 4. Calculate the control limits for the I chart.

$$\text{UCL}(X) = \text{Xbar} + (2.66)(\text{MRbar})$$
$$= 103.3 + (2.66)(18.556)$$
$$= 103.3 + 49.36$$
$$= 152.66$$

$$
\begin{aligned}
\text{LCL}(X) &= \text{Xbar} - (2.66)(\text{MRbar}) \\
&= 103.3 - (2.66)(18.556) \\
&= 103.3 - 49.36 \\
&= 53.94
\end{aligned}
$$

Note: If this LCL(X) turns out to be negative and the data could not be negative (as in this case since a time period cannot be less than zero), the lower control limit would be specified as 0.

Step 5. Plot the centerline and the control limits, as illustrated in Figure 4.1.

Step 6. Plot the individual measurements as illustrated in Figure 4.1. □

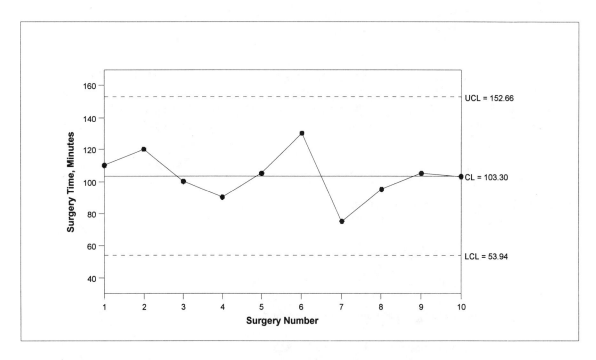

Figure 4.1 Individual Chart on the First Ten Lap Chole Survey Times in Minutes

Process Improvement: Test for Results That May Suggest Nonrandom Influence

To interpret the I chart in Figure 4.1, you must seek results that suggest the existence of nonrandom influence. Tests for this purpose have been developed, such as a single point outside the 3-sigma limits (Shewhart's original "Criterion I" [1931]), plus additional "sensitizing tests." These tests were intended for use with 3-sigma limits on a time-ordered control chart of, say, 24 to 30 points and are discussed in detail in Appendix 1. Nelson's tests [1984, 1985] are discussed here in brief.

The area on the control chart between the centerline and each control limit is divided into three equal zones, zone C between the centerline and 1-sigma, zone B between 1-sigma and 2-sigma, and zone A between 2-sigma and 3-sigma. The three zones above the centerline are illustrated in Figure 4.2. There are three such zones below the centerline as well. Considering only one side of the centerline at a time, Nelson's tests are as follows:

Test 1.   One point beyond zone A, that is, 3-sigma (Figure 4.3)
Test 2.   Nine points in a row in zone C or beyond, that is, on one side of the centerline (Figure 4.4)
Test 3.   Six points in a row (including endpoints) steadily increasing or decreasing (Figure 4.5)
Test 4.   Fourteen points in a row alternating up and down (Figure 4.6)
Test 5.   Two out of three points in a row in zone A or beyond, that is, outside the same 2-sigma (Figure 4.7)
Test 6.   Four out of five points in a row in zone B or beyond, that is, outside the same 1-sigma (Figure 4.8)
Test 7.   Fifteen points in a row in zone C (above and below centerline), that is, within 1-sigma (Figure 4.9)
Test 8.   Eight points in a row on both sides of the centerline with none in zone C, that is, none within 1-sigma (Figure 4.10)

Most current software packages tend to follow the Western Electric/AT&T [1956] and/or the Nelson [1984, 1985] tests.

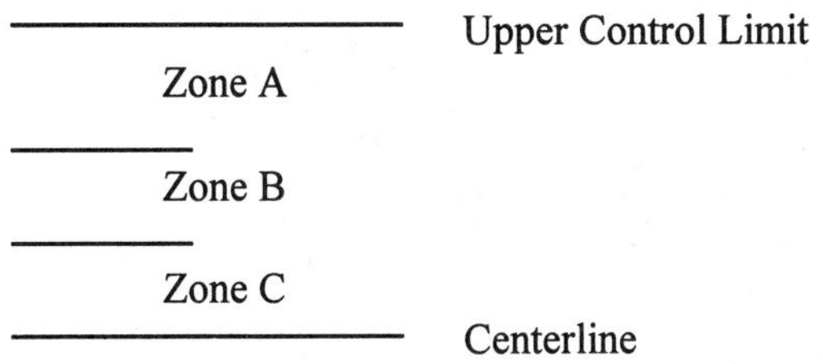

Figure 4.2 Zones on the Control Chart

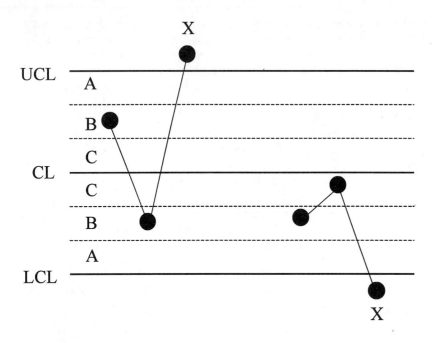

Figure 4.3 One Point beyond Zone A (Nelson's Test 1)

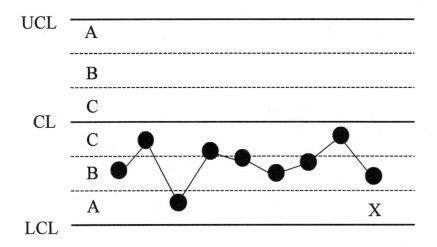

Figure 4.4 Nine Points in a Row in Zone C or Beyond (Nelson's Test 2)

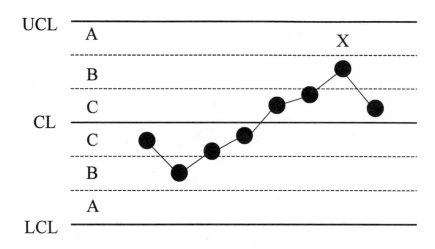

Figure 4.5 Six Points in a Row (Including Endpoints) Steadily Increasing or Decreasing (Nelson's Test 3)

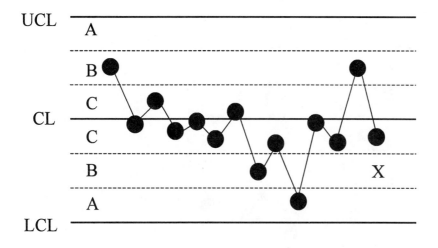

Figure 4.6 Fourteen Points in a Row Alternating Up and Down (Nelson's Test 4)

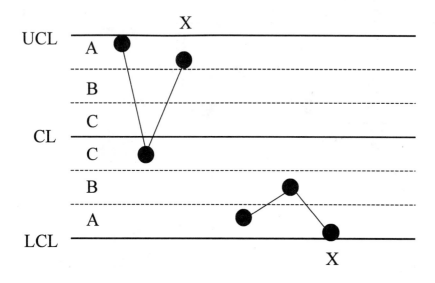

Figure 4.7 Two Out of Three Points in a Row in Zone A or Beyond (Nelson's Test 5)

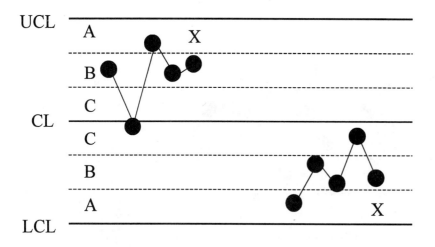

Figure 4.8 Four Out of Five Points in a Row in Zone B or Beyond (Nelson's Test 6)

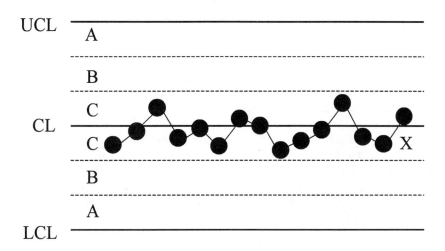

Figure 4.9 Fifteen Points in a Row in Zone C (above and below the Centerline) (Nelson's Test 7)

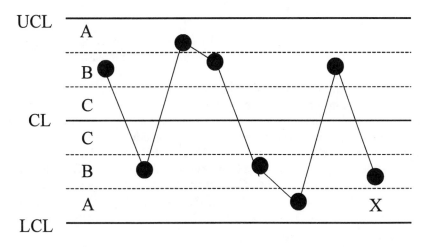

Figure 4.10 Eight Points in a Row on Both Sides of the Centerline with None in Zone C (Nelson's Test 8)

Even with just the first test, the α-risk is quite high with just 25 plotted points. (If the data are normally distributed with standard given, the probability of a point being inside 3-sigma limits is 0.9973. The probability of all 25 points being inside 3-sigma is then $0.9973^{25} = 0.9346$, and the probability of one or more points outside of 3-sigma is then $1 - 0.9346 = 0.0654$.) With no standard given, Hart, Hart, and Philip [1992] found that for an Xbar and s chart with 25 subgroups, the total α-risk for one or more outages on one or both charts runs from 0.11 (for $n = 20$) to 0.27 (for $n = 2$).

69

As more tests are used, the total α-risk increases quite dramatically. Hilliard and Lasater [1966, p. 56] made a study of the use of multiple criteria:

> A large number of control charts, no standard given, for means and ranges of 25 samples of size four were analyzed using the limits test, the length of runs test, and the number of runs test. Results revealed that the probability of one or more false indications of nonrandomness on either the chart for means, the chart for ranges, or both is approximately 0.25 when all three tests are used on each chart.

Due to these large α-risks, it is recommended here that for the evaluation of a process (i.e., the judgment of whether it is a common-cause system), only the 3-sigma criteria be used. However, when the objective is to actively improve the process, the false alarm risk is of little importance. You are not as concerned about investigating the process to find special causes of variation when none exist as you are about finding them as soon as possible if they do exist. Hence, for process improvement, any or all criteria may be used. Many wild-goose chases may be undertaken, but if the process really has special-cause variation, it is important that these be discovered and dealt with.

From the I chart in Figure 4.1, no results can be found suggesting the existence of nonrandom influence, so it will be necessary to look further for guidance on how best to implement process improvement efforts.

The I Chart to Judge Whether a State of Control May Be Inferred

The use of the I chart for process improvement was discussed above. A second use for the I chart is to judge whether a state of control may be inferred. This knowledge may be important for several reasons:

1. If a state of control exists, future performance may be predicted.
2. Whether a state of control may be inferred may be important information in its own right, either to quality personnel within the organization or to an outside quality agency.
3. The process should be in control before a process capability is inferred (see Chapter 6).
4. It is beneficial for the process to be in control before using it for standard values.
5. Perhaps most important, all processes cannot be improved at the same time. Determining which processes are out of control may be an important part of a screening activity to determine where to use limited quality improvement resources.

Standard Given

The use of standard given for an I chart is similar to that with a run chart. However, in addition to having a standard value for the centerline, standard values are also used for the control limits. It may be observed that this is equivalent to having standard values for Xbar and MRbar, or to having standard values for the process average (μ) and the process standard deviation (σ). Standard values may come from

1. data collected earlier on this process (for ongoing process control or improvement)
2. data collected from another application (for comparison)
3. an outside standard (for comparison)

When I chart control limits from process data are to be used as standard values, it is beneficial for the process to be in control, and the limits should be calculated from no less than 24 data points. (Fewer points are used as trial limits discussed earlier in this chapter.) In this case the centerline and control limits are projected into the future. New data are plotted against those standard values as they are collected. Ongoing limits (usually 3-sigma) should be updated whenever there has been a known change in the process, or the data show that the process has changed, and so on. To judge that a state of control may be inferred, 24 successive points falling within control limits are recommended. On the other hand, lack of control may be evidenced by one or more failures to meet control limits with a much smaller number of subgroups.

When attempting to interpret a historical I chart, there may be evidence of special-cause variation but no way (at such a late date) to identify the nature of the special cause. The probability of identifying a special cause is greatly increased if the control chart limits are carried forward and used as standard values [ANSI Standard Z1.2, 1958, 1975; Bicking and Gryna, 1974] for this prospective (as opposed to retrospective) use of the control chart, where data are plotted as soon as they are collected.

EXAMPLE 4.2

To illustrate the technique of standard given, an I chart will be made with no standard given on the 40 surgery times in Table 4.2. These limits are then used as standard values for subsequent observations. To determine whether the 40 observations may be accepted as near-normal, the histogram and probability plots are shown in Figures 4.11 and 4.12. Again, the probability plot indicates that a straight line would be a satisfactory fit. (The histogram only indicates that the data are not badly skewed.) The data are accepted as near-normal. The I chart may therefore properly be made and is shown in Figure 4.13. The reader may verify the calculation of the centerline and control limit values shown in Figure 4.13:

$$CL = Xbar = 90.1$$
$$UCL(X) = 157.6$$
$$LCL(X) = 22.6$$

Assume that the next surgery time turned out to be 160. Should you infer that this new surgery was part of the original system, or is this surgery time outside the control limits? The new observation is shown on Figure 4.14, an I chart using the centerline and control limits from Figure 4.13 as standard values.

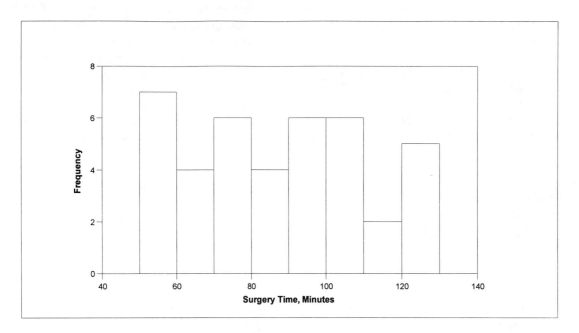

Figure 4.11 Histogram of 40 Surgery Times

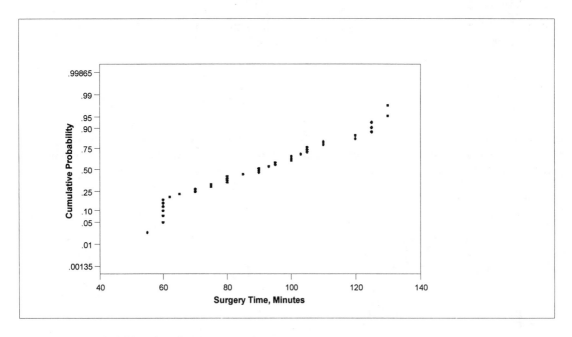

Figure 4.12 Probability Plot of 40 Surgery Times

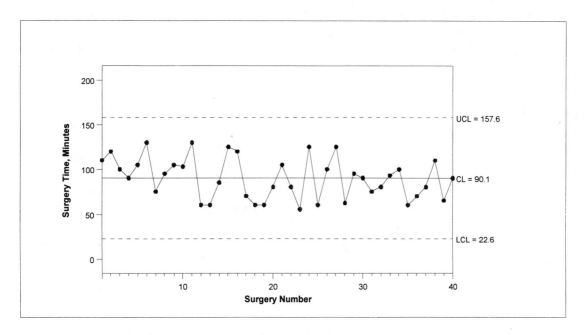

Figure 4.13 I Chart of 40 Surgery Times

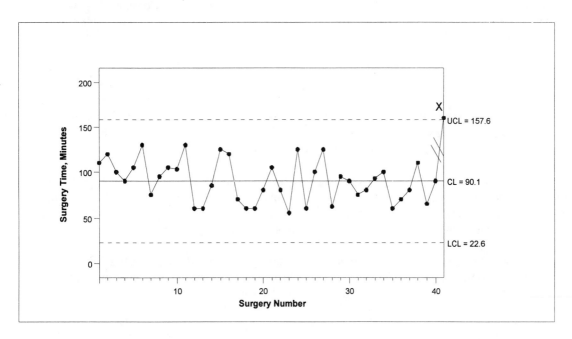

Figure 4.14 I Chart with First 40 Surgeries as "Standard Given"

Since 160 is above the upper control limit, it is concluded that special-cause variation exists. It would be necessary to investigate the unusual nature of this surgery. Perhaps it had more complications. Is this something that could have been avoided? Or, at least, could it have been predicted beforehand? If so, this would help with future scheduling. □

EXAMPLE 4.3

A new surgeon has joined the staff and has started to do lap choles. In order to better schedule the operating room for this new surgeon, it is desirable to determine whether his operating times are statistically equivalent to the present surgeons. To illustrate the concept again with a short data set, suppose that the data from Table 4.2 all came from the present surgeon, surgeon A. The new surgeon, surgeon B, has just completed two surgeries of durations 120 and 165 minutes. Is there evidence of special-cause variation between the two surgeons?

Using surgeon A's results as the standard values,

Xbar = 90.1
$UCL(X) = 157.6$
$LCL(X) = 22.6$

Since surgeon B's second surgery time of 165 falls above the upper control limit, you must infer that special-cause variation does exist. Perhaps more time needs to be scheduled for surgeon B. Note that two new observations were sufficient to indicate lack of control with standard given. However, 24 new observations would have been required to have sufficient statistical evidence to infer that a state of control with standard given did exist. □

The Interrupted I Chart

As with the interrupted run chart, interruptions in the process or data acquisition should be emphasized on the interrupted I chart by selective interruption of the connecting lines. Where applicable, the use of annotation or special plotting symbols may also be useful.

With the interrupted I chart, special care must be taken in calculating MRbar. The moving range value across the interruption should not be calculated. The moving range calculated across a process interruption is often inflated because of a change in the level of the process during the interruption. To judge whether a state of control may be inferred, there should be enough observations to yield at least 23 valid MR values.

This is illustrated in Example 4.4. (Some computer software packages handle this situation by putting a "missing value" symbol in the data list to denote the interruption. Caution: some software packages do not make the interrupted I chart correctly, e.g., they calculate the MR values over process interruptions.)

EXAMPLE 4.4

For instance, suppose that the first six observations from Example 4.1 are consecutive observations from the first week in January, then no data were collected again until March. Note the single missing X value (to denote an interruption) and the two resulting places where calculated MR values would be "not applicable" (NA).

The X and MR values are

X	110	120	100	90	105	130	?	75	95	105	103
MR	NA	10	20	10	15	25	NA	NA	20	10	2

where "?" is the missing value symbol, in this case denoting an interruption.

Then

Xbar = 1033/10 = 103.3 (as before)
MRbar = 112/8 = 14 (Note there are only 8 valid MR values.)

Here there are 10 points plotted on the I chart and 8 MR values.

The control limits are calculated from these values as

$$\text{UCL}(X) = \text{Xbar} + (2.66)(\text{MRbar})$$
$$= 103.3 + (2.66)(14)$$
$$= 103.3 + 37.2$$
$$= 140.5$$
$$\text{LCL}(X) = \text{Xbar} - (2.66)(\text{MRbar})$$
$$= 103.3 - (2.66)(14)$$
$$= 103.3 - 37.2$$
$$= 66.1$$

Note the differences between these limits (shown in Figure 4.15) and those in Figure 4.1. Again, the process appears to have only common-cause variation. □

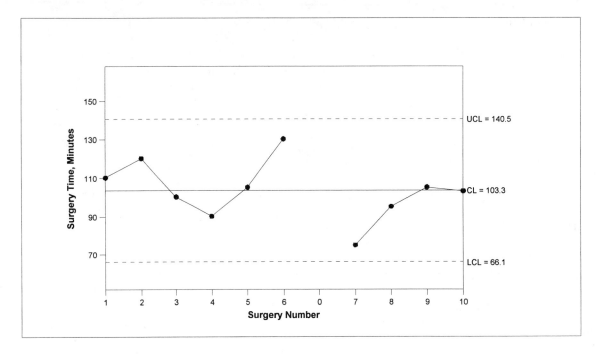

Figure 4.15 Interrupted I Chart

The I Chart with Other Types of Data

The use of the I chart that has been considered here is its primary application, for individual observations with variables data. The I chart may sometimes also be used to advantage with either variables or attribute data to plot any of the subgroup statistics in time order.

Chapter Summary

- The I chart is often the best chart for time-ordered data.

- The control limits for an I chart assume a *near-normal* distribution.

- Control limits for an I chart are calculated with 3σ (or T-sigma) limits.

- The process standard deviation (σ) is estimated from MRbar (i.e., from "within-subgroup" variation).

- Formulas:
  $\text{UCL}(X) = \text{Xbar} + (2.66)(\text{MRbar})$
  $\text{LCL}(X) = \text{Xbar} - (2.66)(\text{MRbar})$ (LCL reset to 0 if LCL is negative, but the data cannot be negative).

- The MR chart is not recommended.

- As in all time-ordered control charts, the I chart is most effective for process improvement if the data are from a single stream.

- At least 24 observations are recommended for formal evaluation for stability/predictability, but fewer may be used to start process improvement efforts.

- An interrupted I chart is made to accommodate an interruption in the process and/or sampling. It does not calculate MR values across interruptions.

- With no standard given, the control limits are calculated from the data.

- With standard given, standard values may come from

  1. data collected earlier on this process (for ongoing process control and improvement)
  2. data collected from another application (for comparison)
  3. an outside standard (for comparison)

- For ongoing process improvement, the standard values usually come from data collected earlier on the same process. The centerline and control limits are projected into the future and new data are plotted in real time against those standard values as they are collected.

Problems

1. Make an I chart using only the first 12 values in Table 4.2. Is there evidence of special-cause variation? Is there sufficient data for this chart?

2. Make an I chart using only the first 12 values in Table 4.2 and a standard value of 80 for the centerline. Is there evidence of special-cause variation? Is there sufficient data for this chart?

3. Make an I chart on the following 30 time-ordered surgery times (in minutes) for lap choles by surgeon A. What do you see from the data?
   65 75 75 60 165 75 85 85 80 95 65 65 85 68 190 120 105 115 58 70 80 75 65 75 90 75 65 50 80 75

4. Make an I chart on the following time-ordered surgery times (in minutes) for lap choles by a particular surgeon. What do you see from the data?
   250 100 75 75 70 155 65 60 75 70 88 90 115 90 75 75 120 125 85 80 105 95 80 85 150 100 85 75 80 80

5. Make an I chart on the following time-ordered surgery times (in minutes) for lap choles by a particular surgeon. What do you see from the data?
   80 115 75 95 130 125 105 70 72 95 90 120 75 90 85 90 80 115 80 65 70 185 75 90 120 130 65 65 115 65

6. Make an I chart on the following number of days between surgical site infections for a given surgery. What do you see from the data?
    4 8 3 2 5 7 6 11 2 7 0 1 13 22 15 2 11 3 5 2 4 2 3 2 2 8 1 6 4 1 0 11 0 4 6 19 1 16
    0 2 20 1 7 4 2 5 6 2 1 10 3 17 3 10 30 9 5 1 3 25 8 26 34 7 5 2 4 21

7. Make an interrupted I chart from the following data. (A ? means an interruption.) What do you conclude?
    120 110 110 95 100 ? 135 75 95 105 110 130 ? 60 65 85 125 120 75 70 60 80

8. Given the following control chart for process improvement, cite all evidence of nonrandom variation.

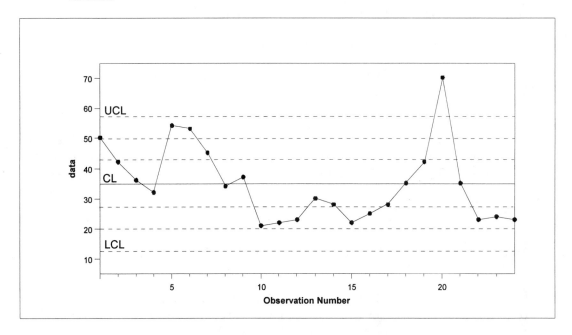

9. Make an interrupted I chart from the following data. (A ? means an interruption.) What do you conclude?
    105 85 75 115 60 ? 100 120 ? 60 95 90 70 80 95 100 ? 65 70 80 110 65 100

Computer Supplement (Statit)

EXAMPLE 1

Individual Chart:
With data set: sem1.wrk
>   [I] button (stands for Individual)
>   >   Data Variable: (click on ▶, Surgery_minutes, Done)
>   >   X Labels variable: (click on ▶, Date, Done)
>   >   Individual Chart Control limits:
>   >   >   (Leave all on Auto unless the LCL turns out to be negative and if the data cannot
>   >   >   be negative, then specify the LCL to be 0. To specify the LCL to be 0,
>   >   >   >   at Individual Control Chart Limits, Lower:
>   >   >   >   >   click on ▼, Specify, and type in *0* in the box on the right.)
>   >   X-Axis label frequency: (type in *1*)
>   >   Chart title: (type in *Surgery Times*)
>   >   Sub title: (type in *minutes*)
>   >   OK ☐

QC Preferences

QC preferences allow you to get additional information on the chart. Set these before making any charts and they will apply to all control charts until they are changed again.

EXAMPLE 2

>   Edit > Preferences > QC . . .
>   >   Limit line text: (click on ▼, then click on
>   >   >   Value and Abb.)
>   >   √ at Display data into table below chart
>   >   √ at Display summary information at bottom of chart
>   >   Number of digits after decimal: (type in *1*)
>   >   OK
Go back and make the I chart again.
>   [I] button
>   OK (if all the choices are still there; otherwise, redo previous example.) ☐

EXAMPLE 3

Individual Chart Using Standard Given:
Projecting old limits into the future: Work with data set: sem1.wrk.

Add a new case (row) of data: 09-May-01   160

Recall that on the I chart, LCL(X) = 54.0, Xbar = 103.3, and UCL(X) = 152.6. Make an I chart using these limits as standard given to see if the new data appear to be part of the old process.

[I] button
Data Variable: (click on ▶, Surgery_minutes, Done)
X Labels variable: (click on ▶, Date, Done)
Individual Chart Control limits:
Upper: (click on ▼ then Fixed and type in *152.6*)
Center:(click on ▼ then Fixed and type in *103.3*)
Lower: (click on ▼ then Fixed and type in *54.0*)
X-Axis label frequency: (type in *1*)
Chart title: (type in *Surgery Times*)
Sub title: (type in *using standard given*)
OK ☐

EXAMPLE 4

Using limits from another application:
Clear the old data (File > Clear data). Work with data set: sem1.wrk.

Let's add another variable (column), Surgeon, to sem1.wrk, so the data looks as follows:

	Date	Surgery_minutes	Surgeon
1	04-Jan-2001	110	A
2	05-Jan-2001	120	A
3	09-Jan-2001	100	A
4	23-Jan-2001	90	A
5	25-Jan-2001	105	A
6	31-Jan-2001	130	B
7	14-Feb-2001	75	B
8	12-Mar-2001	95	B
9	04-May-2001	105	B
10	05-May-2001	103	B

To add the new variable:
Be in the Data Window of sem1.wrk, then choose:
Edit > New Variable
Variable Name: type in *Surgeon*

Data Type: √ String
OK

To save the data set:

Click on a cell, then type in the rest of the data. Save the data as sem2.wrk

> File > Save Data File . . .
>> Save in: click on ▶, then 3 1/2 Floppy (A:) if desired
>> File name: type in *sem2*
>> Save

To delete data:

We wish to get the control limits only from surgeon A. To delete the data from surgeon B:

> Click and drag on 6-10 in the left column to select cases 6-10, and release.
> Right click
> Click on Delete. Warning: "This operation cannot be undone"
> Click on Continue

Make the I chart on surgeon A:

[I] button

> Data Variable: (click on ▶, Surgery_minutes, Done)
> X Labels variable: (click on ▶, Date, Done)
> Individual Chart Control limits:
>> all should be on Auto
> X-Axis label frequency: (type in *1*)
> Chart title: (type in *Surgery Times, Surgeon A*)
> Sub title: (type in *no standard given*)
> OK

Note that UCL(X) = 141.6, Xbar = 105.0, and LCL(X) = 68.4.

Reload sem2.wrk:

File > Open Data File . . .

>> Message: "Do you want to save changes . . . ?"
>> Click on No
>> Look in: (click on ▼, then on 3 1/2 Floppy (A:) if desired
>> File name: click on sem2.wrk
>> Open

Make the I chart on all the data using surgeon A as standard given:

[I] button

> Data Variable: (click on ▶, Surgery_minutes, Done)
> X Labels variable: (click on ▶, Date, Done)
> Individual Chart Control limits:
>> Upper: (click on ▼ then Fixed and type in *141.6*)
>> Center:(click on ▼ then Fixed and type in *105.0*)
>> Lower: (click on ▼ then Fixed and type in *68.4*)

X-Axis label frequency: (type in *1*)
Chart title: (type in *Surgery Times*)
Sub title: (type in *using Surgeon A as standard given*)
OK ☐

Interrupted Individual Chart

Statit uses a ? (the missing value symbol) to denote an interruption in the process.

EXAMPLE 5

Open sem3.wrk.

	Date	Surgery_minutes
1	1-4-01	110
2	1-5-01	120
3	1-9-01	100
4	1-23-01	90
5	1-25-01	105
6	1-31-01	130
7	2-1-01	?
8	2-14-01	75
9	3-12-01	95
10	5-4-01	105
11	5-5-01	103

[I] button
    Data Variable: (click on ▶, Surgery_minutes, Done)
    X Labels variable: (click on ▶, Date, Done)
    Individual Chart Control limits:
        (leave all on Auto)
    X-Axis label frequency: (type in *1*)
    Chart title: (type in *Surgery Times*)
    Sub title: (type in *minutes*)
    OK ☐

Computer Supplement (Minitab)

EXAMPLE 1

Individual Chart:
With data set: sem1

        Stat > Control Charts > Individuals . . .
                Variable: (click on C2 Surgery_minutes, Select)
                OK ☐

EXAMPLE 2

Individual Chart Using Standard Given:
Projecting old limits into the future: Work with data set: sem1

Add a new case (row) of data: 5-9-01   160

Recall that on the I chart, LCL(X) = 54.0, Xbar = 103.3, and UCL(X) = 152.6. Make an I chart using these limits as standard given to see if the new data appear to be part of the old process.

Note: For the I chart, σ may be estimated by: (UCL − Xbar)/3.
So here: (152.6 − 103.3)/3 = 16.43

        Stat > Control Charts > Individuals . . .
                Variable: (click on C2 Surgery_minutes, Select)
                Historical mean: (type in *103.3*)
                Historical sigma: (type in *16.43*)
                Annotation (click on ▼, Title . . .)
                      Title: (type in *Standard Given*)
                      OK
                OK ☐

EXAMPLE 3

Using limits from another application:
Clear the old data (File > Close Worksheet, No). Reopen data set sem1.

Type in another variable (column), Surgeon, to sem1.wrk so the data looks as follows:

	Date	Surgery_minutes	Surgeon
1	04-Jan-2001	110	1
2	05-Jan-2001	120	1
3	09-Jan-2001	100	1
4	23-Jan-2001	90	1
5	25-Jan-2001	105	1
6	31-Jan-2001	130	2
7	14-Feb-2001	75	2
8	12-Mar-2001	95	2
9	04-May-2001	105	2
10	05-May-2001	103	2

To save the data set:
> File > Save Worksheet As . . .
>> Save in: click on ▶, then 3 1/2 Floppy (A:) if desired
>> File name: type in *sem2*
>> Save

To delete data:
We wish to get the control limits only from surgeon 1. To delete the data from surgeon 2:
> Click and drag on 6 – 10 in the left column to select cases 6 – 10, and release.
> Right click
> Click on Delete Cells

Make the I chart on surgeon 1:
> Stat > Control Charts > Individuals ...
>> Variable: (click on C2 Surgery_minutes, Select)
>> Annotation (click on ▼, Title . . .)
>>> Title: (type in *Surgeon 1*)
>>> OK
>> OK

Note that UCL(X) = 141.6, Xbar = 105.0, and LCL(X) = 68.4,
and σ may be estimated by (UCL – Xbar)/3.
So here: (141.6 – 105)/3 = 12.2

Reload sem2. Make the I chart on all the data using surgeon 1 as standard given:
> Stat > Control Charts > Individuals . . .
>> Variable: (click on C2 Surgery_minutes, Select)
>> Historical mean: (type in *105*)
>> Historical sigma: (type in *12.2*)
>> Annotation (click on ▼, Title . . .)
>>> Title: (type in *Limits from Surgeon 1*)
>>> OK
>> OK □

EXAMPLE 4

Interrupted I Chart:
Open sem3.xls. (*Note*: format of dates may be different.) Minitab uses an * (the missing value symbol) to denote an interruption in the process.

	Date	Surgery_minutes
1	1-4-01	110
2	1-5-01	120
3	1-9-01	100
4	1-23-01	90
5	1-25-01	105
6	1-31-01	130
7		*
8	2-14-01	75
9	3-12-01	95
10	5-4-01	105
11	5-5-01	103

Stat > Control Charts > Individuals . . .
>   Variable: (click on C2 Surgery_minutes, Select)
>   OK ☐

Case Study 4.1 Monitoring Blood Pressure Measurements for an Individual Patient

This case study was developed with the assistance of Richard Stanula of Pleasant Valley, Wisconsin, an advocate of statistical process control and a friend for 20 years.

Background

Many times, the opportunity to learn from patient data is lost because the data are not monitored graphically over time. Thanks to Larry V. Staker, the following quotations are offered:

Berwick's Rule: If you pick something to measure and follow it over time, *good things will happen!*

Don Berwick

and

Staker's Corollary: If you pick something to measure and follow it over time, then do something different while continuing to follow it over time, *great things will happen!*

Larry V. Staker

In this case study, the blood pressure of a male patient in his forties was followed intermittently over a five-and-a-half-year period. During that time period, two important things changed: the level of exercise activity and the stress level of the job.

The Data

(Data are in file ZRSBP.) Table 1 shows the date and the systolic and diastolic blood pressure measurements. The question marks in Table 1 are the software missing value symbols, which are used here to indicate interruptions to the process. The first column is called "Line #" to emphasize that in the 24 lines of data, there are only 21 pairs of blood pressure measurements.

Case Study 4.1 Table 1 Blood Pressure Measurements

Line #	Date	Systolic	Diastolic
1	15-Dec-02	120	84
2	18-Jan-03	120	80
3	14-Apr-03	130	92
4	08-May-03	118	80
5	12-Apr-04	120	90
6	10-Nov-04	128	80
7	05-Jun-05	120	80
8	?	?	?
9	14-Oct-05	144	100
10	08-Nov-05	150	110
11	?	?	?
12	04-Dec-05	130	80
13	20-Jan-06	122	94
14	22-Feb-06	118	78
15	05-Apr-06	128	78
16	03-May-06	118	80
17	?	?	?
18	21-Feb-07	110	78
19	03-Jul-07	102	78
20	21-Dec-07	118	88
21	10-Jan-08	104	70
22	17-Feb-08	120	80
23	13-Jun-08	110	70
24	15-Jul-08	110	84

It is important to note that the data should include more than just the measurements themselves. There is critically important data that are *not* given in Table 1: the conditions under which the numerical data were gathered.

The interruptions to the process divide the data into four time eras. In the first three time eras, the patient had a moderate exercise level, approximately equivalent to 0.5 hours of walking per day. In the fourth era, the patient walked 1.5 hours per day. The patient had the same job and about the same stress levels in all except the second era (where there are only two pairs of measurements). In the second time era, the patient was temporarily on a job that he found to be very stressful.

Analysis, Results, and Interpretation

The differing conditions of the four eras suggest that the data may be stratified, with the first and third eras defining the baseline or ordinary process. It might be expected that the second era would yield systematically higher blood pressures with the fourth era systematically lower.

All data analysis efforts, whether in classical statistics or in SPC, focus on the assessment of whether the observed variation may reasonably be attributed to random chance. The first step in process improvement must be to attempt to discover all special causes of variation. This is often done most effectively by studying the tabular data on paper to seek patterns in the "extreme values" that suggest nonrandom influence.

Systematically high blood pressures may be expected in era 2. Starting with the systolic measurements, the reader is encouraged to circle the highest systolic value. The value of 150 is in era 2, as expected. The probability that this is due to chance is 2/21, since 2 of the total of 21 values are in era 2. Circling the next highest systolic value, it is found to be 144—also in era 2. The probability of this second highest value occurring in era 2 by chance is 1/20, since only 1 of the remaining 20 values is in era 2. If the two outcomes were independent, the probability of the two highest values both falling in era 2 by chance is the product of $(2/21)(1/20) = 0.005$. The diastolic results are equivalent. These results cannot reasonably be attributed to chance. As anticipated, the blood pressure values are stratified by era.

Similarly, systematically low blood pressures are expected in era 4. Putting triangles around the lowest systolic values, it is seen that the lowest systolic value of 102 is one of the seven values in era 4. You may continue with putting triangles around the lowest remaining values until the five lowest have been identified, discovering that all five occur in era 4. The probability of this happening by chance is again very low. The reader may verify that the two lowest diastolic measurements fall in era 4. Again, the results cannot reasonably be attributed to chance and the blood pressure results appear stratified by era. Note that the similarity of the systolic and diastolic results adds additional credibility to this statistical analysis.

Since the data are stratified by era, even though the mean of the data can be calculated, it has no meaning. The same is true of any type of estimate that might be made for the standard deviation. A control chart made on these data will have control limits that do not describe the real process. To make a control chart, you assume a state of control. *After* making the control chart, you may conclude whether or not that assumption was correct.

A second method of preliminary analysis is the use of an interrupted run chart. Figures 1 and 2 are interrupted run charts for the systolic and diastolic measurements. Recall that the run chart is simply a record of the data over time and makes no assumption about the normality of the data. The systolic run chart identifies the special nature of eras 2 and 4. The diastolic run chart shows clearly the special nature of era 2, but the indication is less clear with era 4. The interrupted run chart leads to the same conclusion as the analysis of extreme values: the data are stratified. Similar conclusions would be drawn if interrupted control charts had been made on all of the data with no standard given. (See Problems at the end of the case study.)

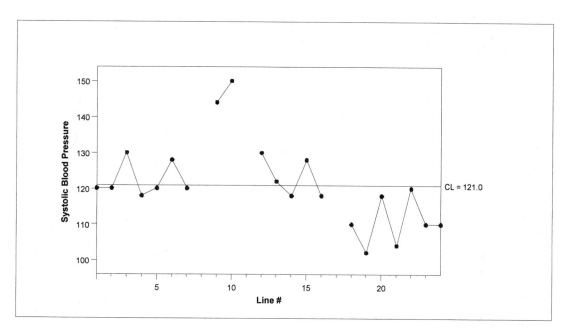

Case Study 4.1 Figure 1 Interrupted Run Chart on Systolic Blood Pressure for the Four Eras

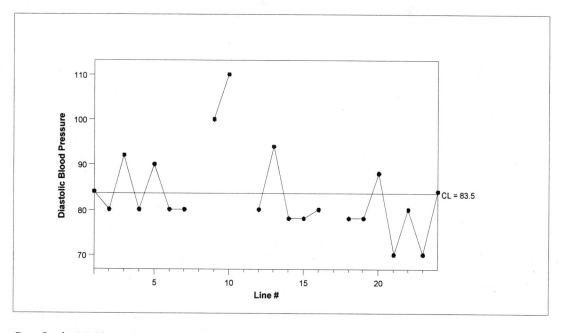

Case Study 4.1 Figure 2 Interrupted Run Chart on Diastolic Blood Pressures for the Four Eras

It is straightforward to construct valid I charts on the baseline processes from eras 1 and 3. These I charts were made by deleting from Table 1 the data lines 17 through 24, then lines 8 through 10. This left 12 systolic values with one interruption and a like set of data for diastolic values. After verifying with probability plots (Figures 3 and 4) that these data may be treated as near-normal, they were used to make I charts for systolic and diastolic blood pressure (Figures 5 and 6). The resulting centerlines and control limits were then used as standard values for the systolic and diastolic control charts shown in Figures 7 and 8.

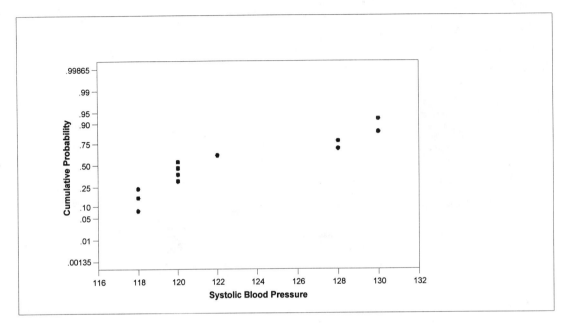

Case Study 4.1 Figure 3 Probability Plot of Systolic Blood Pressure from Eras 1 and 3

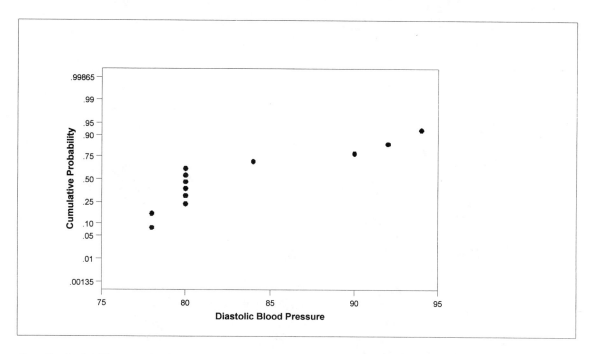

Case Study 4.1 Figure 4 Probability Plot of Diastolic Blood Pressure from Eras 1 and 3

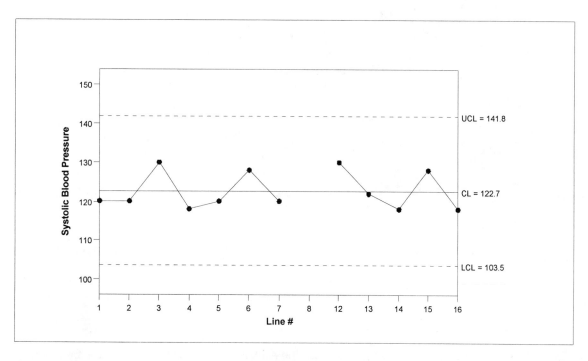

Case Study 4.1 Figure 5 Interrupted Systolic I Chart on Eras 1 and 3

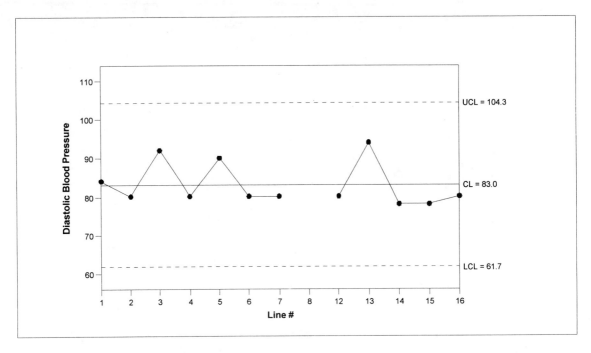

Case Study 4.1 Figure 6 Interrupted Diastolic I Chart on Eras 1 and 3

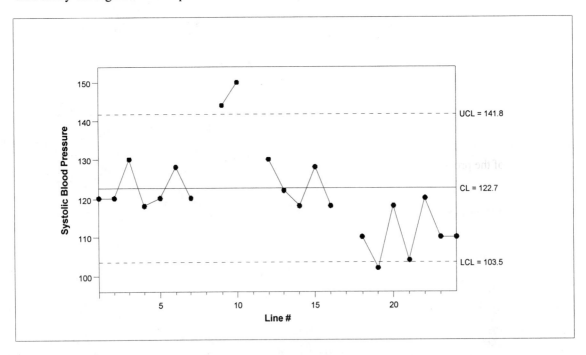

Case Study 4.1 Figure 7 Interrupted Systolic I Chart on All Four Eras Using Standard Values from Eras 1 and 3

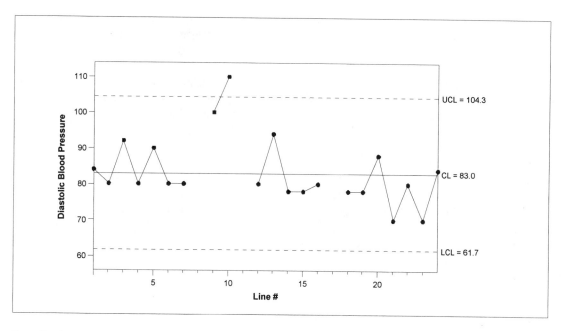

Case Study 4.1 Figure 8 Interrupted Diastolic I Chart on All Four Eras Using Standard Values from Eras 1 and 3

Lessons Learned

An analysis of the patterns in the extreme values of the data or an interrupted run chart is useful to determine whether the data are stratified or are suitable for a control chart. In this case study both methods showed the data to be stratified by era, a "textbook" example of the influences of stress and exercise on blood pressure.

The physician might choose between an I chart for ongoing process control or a run chart for ongoing monitoring of the process, where the run chart has significant advantages.

Management Considerations

All too often decisions are made only on the most recent data point. Instead, the retrospective analysis shown above would make a good point of departure for a prospective, or ongoing, activity to monitor the patient's blood pressure. Such ongoing data analysis efforts have received considerable attention in recent years as a cost-savings initiative for high-cost chronic conditions where patients can monitor their own diet, exercise, and medications and obtain their own data daily for such things as weight, blood pressure, and pulse. A new specialty called "disease management" has emerged, where HMOs, preferred provider organizations, and state Medicaid agencies may have an outside vendor serve as a clearinghouse to receive data over the telephone from patients, monitor the data, and take action as the need arises to prevent costly problems.

Disease management for high-cost chronic conditions is an important healthcare area for the ongoing analyses of time-ordered patient data. However, to include much more than just a few high-profile, high-cost areas, the notion of "ongoing health management" would be appropriate.

Typically, ongoing process control efforts use a run chart on the observations with action limits beyond which observations would be deemed clinically important. This approach misses opportunities for improvement that can be provided only by the control chart method with its statistically derived control limits.

When using a run chart, there is no way to determine whether the process is stable over time, and this knowledge is critical to the management of the process. If the control chart shows that the process is stable, the process capability (the limits of the process beyond which only about one point in a thousand will fall in each tail, see Chapter 6) can be determined and compared with clinically desired limits. The control limits will serve as effective action limits tailored to this individual, rather than action limits that may be about right for the population as a whole. With the proper control limits for this patient's process, notice of important changes to the process can be detected long before they could be found on a run chart, and this early detection may literally be a matter of life and death.

Problems

1. Some books have improperly suggested that "control chart" limits for individuals be set using the classical statistics approach of the mean plus 3 times s for the UCL and the mean minus 3 times s for the LCL.
    a. Why is this wrong?
    b. What values would this give for control limits with the data in Table 1?
    c. Plot these limits on Figures 1 and 2. Would these limits be useful? If not, why not?

2. Make an interrupted I chart for the systolic and the diastolic blood pressure measurements in Table 1. What do you conclude?

Chapter 5 The Xbar and s Chart

The *Xbar and s chart* is a pair of independent control charts, one on the subgroup averages (Xbar values) and one on the subgroup standard deviations (s values). These Xbar and s values are from observations that have been aggregated into subgroups that are as homogeneous as possible (elements within the subgroup are as much alike as possible).

The chart centerlines are X-double-bar and sBar, or

$$\overline{\overline{X}} \text{ and } \overline{s}$$

If the subgroup sizes are constant, these are the averages of the Xbar and the s values. Varying subgroup sizes are treated later in this chapter.

Since sBar is used to calculate the control limits for each chart, it is usually best to interpret the s chart first. The Xbar chart is most believable if the s chart is in control.

The *central limit theorem* states that the distribution of averages of the subgroups will become closer to the normal distribution as the subgroup size gets larger, regardless of the shape of the population distribution. However, the calculation of the within-subgroup estimate of the standard deviation of the averages assumes that the population has a normal distribution. Similarly, the control limits on s assume the data are normally distributed [Duncan, 1986]. However, if the data are near-normal, the control limits will work satisfactorily. In fact, control charts for Xbar are a little more forgiving than are I charts for departures from normality due to the central limit theorem. If the distribution were to be normal, you would expect about 99.73% of the Xbar values to fall within X-double-bar $\pm\, 3\sigma_{Xbar}$ and a high percentage of the s values to fall within sBar $\pm\, 3\sigma_s$. The control limits for Xbar are tighter than those for the individual X values, and some individual values may be expected to fall outside of the Xbar limits.

The Xbar and s Chart with Time-Ordered Data

Recall that the I chart is usually made of a long run of time-ordered data without interruptions (Chapter 4). Although generally less effective than an I chart, a time-ordered Xbar and s chart with small subgroups might be made on the same data. These subgroups usually have equal sizes. For example, while retaining the time order, an Xbar and s chart might be made on the 40 consecutive lap chole surgery times in Table 4.2 after subgrouping as $k = 20$ subgroups of size $n = 2$, or as 10 subgroups of size 4.

Note: For ease of calculation, examples in this book where detailed calculations are shown use only the traditional simplified methods of constant n and 3-sigma limits. However, case studies, where the analysis is done by computer, may have unequal subgroup sizes and use the T-sigma limits from Table 4.1.

Calculations for the Xbar and s Chart: 3-Sigma Control Limits and Constant n

The control limits on the Xbar chart are as follows:
Upper control limit on Xbar:

$$UCL(\overline{X}) = \overline{\overline{X}} + A_3\,\overline{s}$$

Lower control limit on Xbar:

$$LCL(\overline{X}) = \overline{\overline{X}} - A_3\,\overline{s}$$

where A_3 is a factor obtained from Table 5.1 for the given subgroup size, n.

The control limits on the s chart are as follows:
Upper control limit on s:

$$UCL(s) = B_4\,\overline{s}$$

Lower control limit on s:

$$LCL(s) = B_3\,\overline{s}$$

where B_3 and B_4 are factors obtained from Table 5.1 for the given subgroup size, n. Mathematical relations of the formulas for the control limits for the Xbar and s chart may be found in Appendix 3.

The Xbar and s chart with small subgroups may be used as a substitute for the I chart for the analysis of single-stream, near-normal variables data in time order as shown below. However, the Xbar and s chart is much more complicated, is often found to be less powerful [Hart and Hart, 1992], and has other difficulties. Therefore, the I chart is usually more effective. Shewhart recognized that it would be best to plot the individuals rather than to use subgroups for time-ordered data [1931, p. 313 ff].

> The reader may question why the original data were grouped into subsamples of four instead of some other number. A little consideration will show that there is nothing sacred about the number four . . . Obviously, if the cause system is changing, the sample size should be as small as possible so that averages of samples do not mask the changes. Why then do we not use a sample size of unity? The answer is that if we do, we are faced with the difficulty of choosing the standard deviation to be used in control charts. Of course, we might use the standard deviation σ of the entire group of observations but . . . the test in which we would use the standard deviation σ of the whole group of n observations is not so sensitive, in general, as the one in which we divide the data into small subgroups in the order in which they were taken. In fact the sensitivity of the test will increase, in general, with decrease in subsample size . . . there would be some advantage therefore in reducing the subsample size to unity. To do so, however, would obviously defeat our purpose since we could not then obtain an estimate of σ to use in the control charts.

Only because of the problem in estimating the population standard deviation (σ) did Shewhart resort to the use of small subgroups, and this problem has since been solved with the use of the moving range as illustrated in Chapter 4 [Western Electric/AT&T, 1956].

Table 5.1 Factors for Computation of Control Limits

n	A_3	B_3	B_4
2	2.66	0	3.27
3	1.95	0	2.57
4	1.63	0	2.27
5	1.43	0	2.09
6	1.29	0.03	1.97
7	1.18	0.12	1.88
8	1.10	0.18	1.82
9	1.03	0.24	1.76
10	0.98	0.28	1.72
12	0.87	0.35	1.65
15	0.79	0.43	1.57
25	0.60	0.57	1.43
50	0.42	0.70	1.30
∞	(1)	(2)	(3)

$$(1)\ A_3 = \frac{3}{\sqrt{n}}$$

$$(2)\ B_3 = 1 - \frac{3}{\sqrt{2n-2}}$$

$$(3)\ B_4 = 1 + \frac{3}{\sqrt{2n-2}}$$

Tests for Nonrandom Influence

For process improvement when using time-ordered data, the same tests that apply to the I chart apply to the Xbar chart. However, since the distribution of standard deviations is so skewed, the only test that is used on the s chart is one or more points outside the T-sigma control limits.

For judging whether a state of control may be inferred, only the test for one or more points outside the T-sigma limits is applied to both the Xbar and the s chart.

<u>Standard Given</u>

As with the run chart and the I chart, standard values may come from

     1.   data collected earlier on this process (for ongoing process control or improvement)
     2.   data collected from another application (for comparison)
     3.   an outside standard (for comparison)

Example 5.1 illustrates the use the Xbar and s chart with time-ordered data and small subgroups of equal size. As with all of the small numerical examples in this book, this is not intended to be sufficient data for a real application.

EXAMPLE 5.1

To illustrate the calculation of the control limits for the Xbar and s chart with time-ordered data, the 10 observations used for the I chart in Example 4.1 have been subgrouped as $k = 5$ subgroups of size $n = 2$ while retaining their time order. Data subgrouped in this manner might arise, for example, by recording the surgery times for the first two lap choles after 10 a.m. on Tuesday for five consecutive weeks. The values in Table 5.2 are surgery times in minutes. The histogram and probability plots shown in Figures 2.3 and 2.8 are still applicable, and the data are accepted as near-normal.

Table 5.2 First Five Weeks of Lap Chole Surgery Time Data

Week	1	2	3	4	5
First observation	110	100	105	75	105
Second observation	120	90	130	95	103

Step 1. Collect the data.
The first five weeks of data are shown in Table 5.2 as $k = 5$ subgroups of size $n = 2$.

Step 2. Calculate the Xbar and s values from each subgroup.

Week	1	2	3	4	5
Xbar	115.0	95.0	117.5	85.0	104.0
s	7.07	7.07	17.68	14.14	1.41

Step 3. Find X-double-bar ($\overline{\overline{X}}$), the centerline for the Xbar chart, and sBar ($\overline{s}$), the centerline for the s chart.

$\overline{\overline{X}} = (115.0 + 95.0 + 117.5 + 85.0 + 104.0)/5 = 103.3$ (the same as the Xbar in Figure 4.1 since the subgroup sizes were equal)

$\overline{s} = (7.07 + 7.07 + 17.68 + 14.14 + 1.41)/5 = 9.47$

Step 4. Calculate the control limits for each chart.

Note that $n = 2$, so from Table 5.1, $B_4 = 3.27$, $B_3 = 0$, and $A_3 = 2.66$.

For the s chart:
$$UCL(s) = B_4 \overline{s}$$
$$= 3.27(9.47)$$
$$= 30.97$$

$$LCL(s) = B_3 \overline{s}$$
$$= 0$$

For the Xbar chart:
$$UCL(\overline{X}) = \overline{\overline{X}} + A_3 \overline{s}$$
$$= 103.3 + 2.66(9.47)$$
$$= 103.3 + 25.19$$
$$= 128.49$$

$$LCL(\overline{X}) = \overline{\overline{X}} - A_3 \overline{s}$$
$$= 103.3 - 25.19$$
$$= 78.11$$

Step 5. Plot the centerline and the control limits as illustrated in Figure 5.1.

Step 6. Plot the s and Xbar values on their respective charts as illustrated in Figure 5.1. □

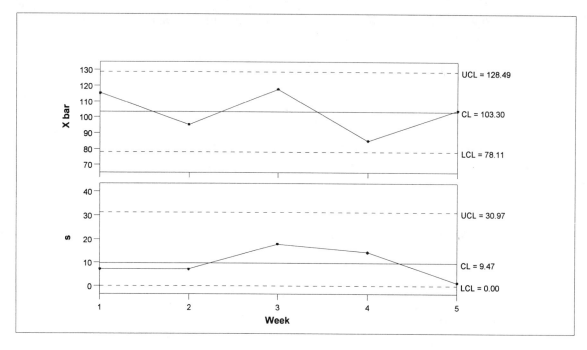

Figure 5.1 Xbar and s Chart for Time-Ordered Data

There is nothing in the Xbar and s chart (Figure 5.1) to suggest nonrandom influence. Note that there appears to be no correlation between the two charts, e.g., they do not "go up and down together" as if they were "in phase." If there were any such correlation, it would indicate that the data were too badly skewed to be near-normal (as will be shown in the case studies).

For the interpretation of the Xbar and s chart in Figure 5.1, you must bear in mind that for process improvement, you seek any results that suggest nonrandom influence where the objective is the prediction of future process performance. This logic is the same as for the I chart for process improvement. From Figure 5.1, it appears that the process may be in control. (However, there is not enough data here to really infer a state of control; 24 subgroups are recommended.)

Some special concerns exist in the use of small subgroups of time-ordered data:

1.   As with the I chart, interruptions in the single stream of time-ordered data may occur because of process changes, because no data were generated, or simply because data were not recorded. With an Xbar and s chart, interruptions in the data and all changes in process conditions must occur *between*, not within, subgroups.

2.   The control limits on the I chart are the natural limits of the individual observations. However, the Xbar values have less variability than do individual observations, so the control limits are narrower than are the control limits for the I chart. *This fact causes many mistakes in drawing inferences about the breadth of the process dispersion!*

100

3.   It is mandatory that the elements of a subgroup be as much alike as possible. A common mistake is to choose subgroups so that there is a systematic variation in the elements. (For example, suppose three measurements are taken each day and the third measurement is always the highest. This artificially inflates the estimate of sigma, resulting in inflated control limits. (This is illustrated in Case Study 5.1 on turnaround time.)

These three items are only some of the problems that are circumvented with the use of an I chart.

For small time-ordered subgroups, an Xbar and R chart (Xbar and Range chart) is sometimes used instead of an Xbar and s chart. The Xbar and R chart has been popular because a subgroup range is easier to calculate by hand than a subgroup standard deviation. However, the Xbar and s chart is statistically a better choice than the Xbar and R chart, and since computers are usually used for the calculations, there is no need to use a shortcut chart. Hence, the Xbar and R chart is not discussed here.

An Xbar and s chart for time-ordered data may also be used for large subgroups, in which case the subgroup sizes often vary. The Xbar and s chart with varying n is too complicated for practical hand calculations but will be illustrated in the computer analyses in the case studies. For charts with varying subgroup sizes, the control limits step up and down, being wider for the smaller subgroups since less certainty exists there. A further complication on the s chart is that there will also be a step in the sBar, the centerline for the s chart, wherever there is a change in subgroup size. The calculation of the control limits for the Xbar and s chart with unequal subgroup sizes are discussed in Appendix 3. The user is cautioned to check the software being used, for some software packages do not handle Xbar and s charts with varying n properly.

The Xbar and s Chart with Rational Subgroups

Up to this point in the book, data analysis has been restricted to time order, where the objective was prediction of the performance of a single stream. Besides analyzing each data stream in time order, it is also essential to make comparisons between different streams—for example, to compare the results of males and females, of various physicians, of day shift and night shift, or of different diagnosis-related groups (DRGs). Comparisons of physicians or of hospitals are often called *profiling*. It may also be necessary to compare "before" and "after," as in the case of a new pathway, or to compare last year to this year, and so on.

Walter Shewhart [1931, p. 299] referred to this type of classification as the use of *rational subgroups*.

> Obviously, the ultimate object is not only to detect trouble but also to find it, and such discovery naturally involves classification. The engineer who is successful in dividing his data initially into *rational* subgroups based upon rational hypotheses is therefore inherently better off in the long run than one who is not thus successful.

Xbar and s charts for rational subgroups will be most effective if made with subgroup sizes of at least 24 in which time order has been retained. This will provide enough data to obtain Xbar and s values, which give reasonable estimates for the population each subgroup represents, and to judge whether each within-subgroup process is stable over time.

The first step in process improvement should be to attempt to discover all special causes of variation. The key to success with the control chart method is being able to anticipate what may be important sources of variation in the process. There will usually be several such sources of variation, and each of these provides its own method of subgrouping the data. For example, there may be potentially important contributions to variation by shift of the practitioners, by gender of the patient, by age of the patient, and so on. Besides analyzing the data in rational subgroups, each single-stream process should be analyzed in time order. The Xbar and s chart for rational subgroups gives an overview of all of the subgroups, while the I chart on the time-ordered X values looks *within* each subgroup. Both are needed.

The discovery of special causes is straightforward with the SPC method. The action to take for improvement is to eliminate these special causes if they are detrimental, make them standard practice if they are beneficial, or at least take them into account in the analysis. Expert knowledge of the healthcare processes is essential in order to know ways that *might* be important to subgroup the data for any particular application. Data should be gathered and the control chart method should be used to analyze the data in each of these ways. There need be no concern about selecting "too many" ways to subgroup the data, for the method is self-correcting. If a particular method of subgrouping the data is not statistically significant (e.g., if there is no significant difference between shifts), the control chart will find no evidence of special-cause variation. But failure to explore the data diligently using rational subgroups will result in lost opportunity for process improvement.

After all special causes have been eliminated so that only common-cause variation remains, a different plan of action is required. In the absence of special-cause variation, process improvement is accomplished only by working to improve the process as a whole, rather than targeting special causes. This will reduce common-cause variation. Taguchi [1981] suggested that "loss" to society can be decreased by setting target values to meet the needs and expectations of customers and then by reducing variation about the target values.

Since SPC identifies the presence or absence of special-cause variation, it indicates the action to be taken for process improvement. It is for this reason that SPC has been called a "road map for improvement." Example 5.2 illustrates the use of the Xbar and s chart with rational subgroups for comparison.

EXAMPLE 5.2

The ten values in Table 5.3 are the same lap chole surgery times used in Examples 4.1 and 5.1, giving both surgeon and anesthesiologist information. Figures 2.3 and 2.8 showed that the data may be accepted as near-normal, so the Xbar and s control chart may be made. To look at the data by surgeon, there will be $k = 2$ subgroups (surgeons) of $n = 5$ surgery times for each. In practice, you would want larger subgroup sizes than the n of 5 shown here.

Both the small subgroups and the use of 3-sigma limits were used to simplify the hand calculations. In practice, 1.5-sigma limits (per Table 4.1) and larger subgroups would be used.

Table 5.3 Lap Chole Surgery Times, Minutes, with Surgeon and Anesthesiologist

Surgery Time Minutes	Surgeon	Anesthesiologist
110	A	D
120	A	C
100	A	C
90	A	D
105	A	C
130	B	C
75	B	D
95	B	D
105	B	C
103	B	C

Step 1. Arrange the data by rational subgroup, by surgeon in this case.
Table 5.4 shows the $k = 2$ subgroups of size $n = 5$.

Table 5.4. Surgery Times, Minutes, for Surgeons A and B

Surgeon	
A	B
110	130
120	75
100	95
90	105
105	103

Step 2. Calculate the Xbar and s values from each subgroup.

Xbar	105.0	101.6
s	11.1803	19.8192

Step 3. Find $\overline{\overline{X}}$, the centerline for the Xbar chart, and $\overline{s}$, the centerline for the s chart.

$$\overline{\overline{X}} = (105.0 + 101.6)/2 = 103.30$$
$$\overline{s} = (11.1803 + 19.8192)/2 = 15.4998$$

Step 4. Calculate the control limits for each chart. Using 3-sigma limits (note that $n = 5$, so from Table 5.1, $B_4 = 2.09$, $B_3 = 0$, and $A_3 = 1.43$):

For the s chart:

$$UCL(s) = B_4 \bar{s}$$
$$= 2.09(15.4998)$$
$$= 32.39$$

$$LCL(s) = B_3 \bar{s}$$
$$= 0$$

For the Xbar chart:

$$UCL(\bar{X}) = \bar{\bar{X}} + A_3 \bar{s}$$
$$= 103.30 + 1.43(15.4998)$$
$$= 103.30 + 22.16$$
$$= 125.46$$

$$LCL(\bar{X}) = \bar{\bar{X}} - A_3 \bar{s}$$
$$= 103.30 - 22.16$$
$$= 81.14$$

Step 5. Plot the centerline and the control limits as illustrated in Figure 5.2.

Step 6. Plot the s and Xbar values on their respective charts as illustrated in Figure 5.2.

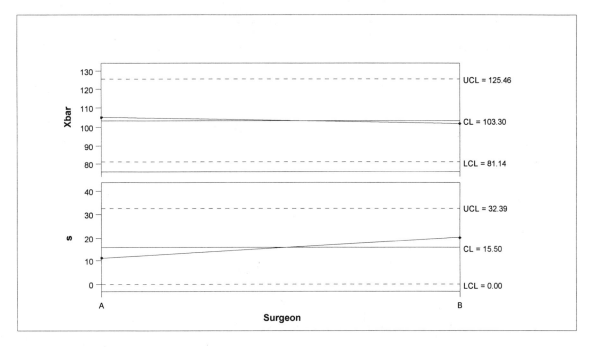

Figure 5.2 Xbar and s Chart for Lap Chole Surgery Times by Surgeon

In Figure 5.2, neither the s chart nor the Xbar chart had points outside the 3-sigma control limits, and by inspection it may be seen that this would also be true with 1.5-sigma limits. This observation may be helpful for process improvement. However, with subgroups of only size 5, there is not enough data to infer that surgeon-to-surgeon variation is in control. If such a conclusion could be inferred, this would mean that the differences between them were reasonably attributed to chance and that in a repeat set of data, their positions on the charts could just as likely be reversed. If one of the surgeons were outside the control limits, this would not imply that one surgeon is doing a better job than the other. It may be that one surgeon has all of the complicated cases; it may be that the anesthesiologist working with the one surgeon takes more time. It is up to the judgment of the surgeons whether the difference is acceptable or whether they might learn from each other. Finding these differences does give an opportunity for discussion and further investigation, which may prove very fruitful if done judiciously.

The data from Table 5.3 may also be analyzed by anesthesiologist.

Step 1.   Arrange the data by rational subgroup, by anesthesiologist in this case.
         Table 5.5 shows the $k = 2$ subgroups of unequal sizes, $n = 6$ and $n = 4$.

Table 5.5 Surgery Times, Minutes, for Anesthesiologists C and D

Anesthesiologist	
C	D
120	110
100	90
105	75
130	95
105	
103	

Step 2.   Calculate the Xbar and s values from each subgroup.

Xbar	110.5	92.5
s	11.81	14.43

Step 3.   Find $\overline{\overline{X}}$, the centerline for the Xbar chart, and $\overline{s}$, the centerline for the s chart.

Because of the unequal-size subgroups, you can't just average the Xbar values to get $\overline{\overline{X}}$. Instead, you get the average of all observations. Here,

$$\overline{\overline{X}} = 1033/10 = 103.3$$

From here on (including the calculation of sBar), the calculations get complicated (see Appendix 3), and we leave them to the computer. The s chart is given in Figure 5.3 and the Xbar chart is given in Figure 5.4. Note how the control limits vary with the subgroup size in both charts. Furthermore, the s chart centerline varies with the subgroup size (although this effect becomes less pronounced as the subgroup sizes become larger).

In practice, rational subgroup sizes usually vary, as seen throughout the case studies. □

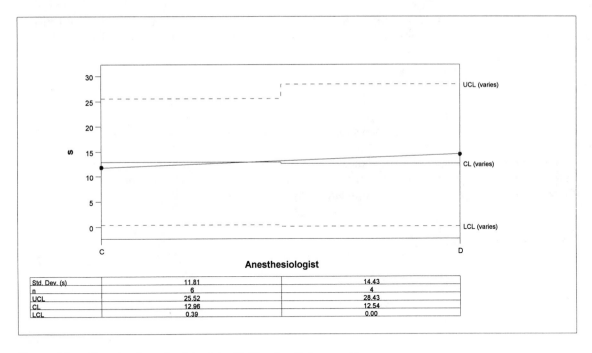

Std. Dev. (s)	11.81	14.43
n	6	4
UCL	25.52	28.43
CL	12.96	12.54
LCL	0.39	0.00

Figure 5.3 s Chart by Anesthesiologist with Varying Subgroup Sizes

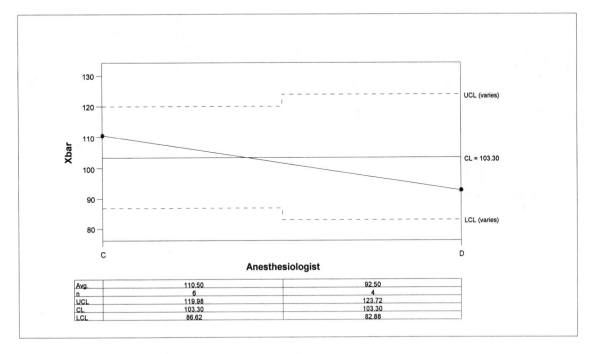

Avg.	110.50	92.50
n	6	4
UCL	119.98	123.72
CL	103.30	103.30
LCL	86.62	82.88

Figure 5.4 Xbar Chart by Anesthesiologist with Varying Subgroup Sizes

107

For rational subgroups that are not time ordered, the Xbar and s chart is used to make comparisons between two or more "populations" for evaluation of whether or not there is special-cause variation between them.

Two methods for comparing two populations have been discussed:

1. The use of standard values from the first population for the evaluation of a second population.
2. The use of an Xbar and s chart with one rational subgroup for each population.

The use of standard given is the better choice when evaluating a new process in real time using a long stable baseline from the first population and perhaps very little data from the second. The Xbar and s chart is used for comparing two or more populations with large subgroups from each.

Tests for Lack of Control

When looking at data by rational subgroups other than time order, only the test for one or more points outside the T-sigma control limits applies. Any "patterns" in the data would be a function of how the subgroups are arranged and would change as the subgroups are rearranged. There is no natural arrangement as with time-ordered subgroups.

How Not to Make Comparisons

The purpose of the control chart here is to answer the question, "Is there evidence of special-cause variation between the subgroups?" The control chart method addresses this question directly. However, other methods are sometimes used to make comparisons using methods that fail to address this issue. Such unsatisfactory methods include bar chart and *box-and-whisker* plots. These methods are not taught here, but Figure 5.5 is a bar chart of the average surgery times of nine surgeons, used in a later case study (6.1), and Figure 5.6 is a series of box-and-whisker plots on the same nine surgeons. There are slight variations, but generally the boxes show the values for the middle 50% of the measurements, that is, 25% fall below the lower end of the box (Q_1) and 25% fall above the upper end (Q_3). The median is shown as a line through the box. A "whisker" is drawn from each end of the box, one to the smallest observation within the median -1.5 $(Q_3 - Q_1)$ and one to the largest observation within the median $+1.5$ $(Q_3 - Q_1)$. Observations beyond these values are generally denoted by a plotted dot. The box-and-whisker plots may be useful to determine the shapes of the various distributions, but note that neither the bar charts nor the box-and-whisker plots tell whether the differences between surgeons are due to special-cause variation or just random chance.

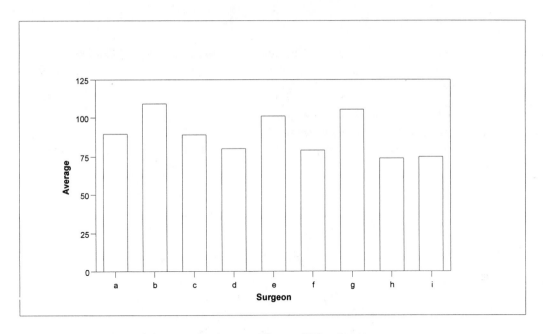

Figure 5.5 Bar Chart of the Average Surgery Times of Nine Surgeons

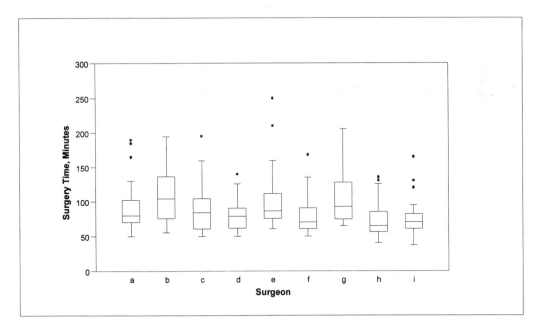

Figure 5.6 Box-and-Whisker Plots of Surgery Times of Nine Surgeons

<u>Representations of Out-of-Control Conditions</u>

It is important for the reader to be able to look at an Xbar and s chart and sketch rough, but useful, distribution curves for each of the subgroups. Consider four applications, each of which has three surgeons (three subgroups) whose surgery times are being compared.

Figure 5.7 shows the sketch of the distribution of the surgery times for each of the three surgeons. The variation within each of the three distributions (the width of each curve) is roughly the same (hence, the s chart would be in control), but their averages are quite different (the Xbar chart would have points outside the control limits).

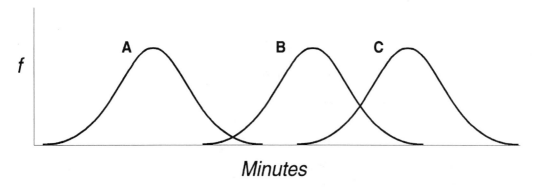

Figure 5.7 Distributions Resulting in the s Chart in Control, Xbar Chart Out of Control

A second application, Figure 5.8, shows that the variation within each of the three distributions is quite different (the s chart would be out of control), but their averages appear to be similar (the Xbar chart would be in control).

110

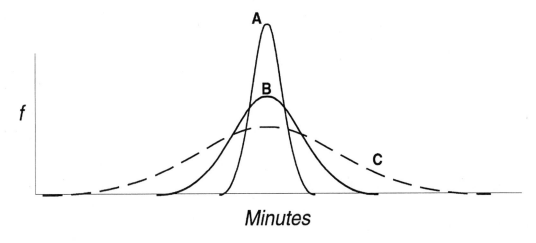

Figure 5.8 Distributions Resulting in the s Chart Out of Control, Xbar Chart in Control

In a third application (Figure 5.9), the variations within the three distributions are quite different (the s chart would be out of control), and their averages appear to be quite different (the Xbar chart would be out of control).

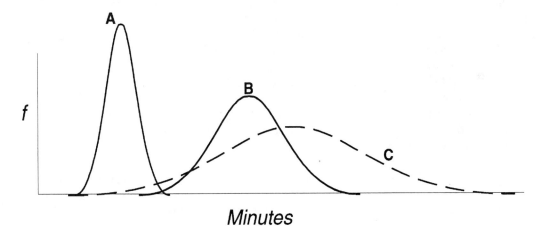

Figure 5.9 Distributions Resulting in the s Chart Out of Control, Xbar Chart Out of Control

A fourth application, Figure 5.10, shows their variation to be similar (the s chart would be in control), and their averages appear to be similar (the Xbar chart would be in control).

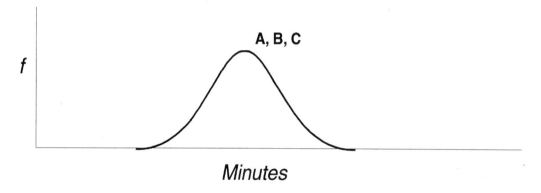

f

A, B, C

Minutes

Figure 5.10 Distributions Resulting in the s Chart in Control, Xbar Chart in Control

Xbar and s Chart Compared to the Analysis of Variance

For the reader familiar with classical statistics, it is interesting to note that the Xbar and s chart makes an analysis similar to that done by the analysis of variance (ANOVA).

> It may be recognized by the reader that the analysis of variance technique . . . will serve as a substitute for an Xbar chart for determining whether a set of past data came from a "controlled process." . . . The analysis of variance is a summary method based on the overall variation among sample means while the control chart for averages is based on the behavior of the individual Xbars. It can happen that no one Xbar will fall outside the limit lines and the analysis of variance will disclose that too many of them are near the limit lines. On the other hand a single point on the Xbar chart might be well outside the control lines and yet because the remaining points are relatively near the center line the total sum of squares among averages will not be large enough for the analysis of variance to indicate lack of control. But in general, the two methods are nearly enough equivalent that both will disclose any clear case of lack of control among averages and when one says control is good the other will too. [Duncan, 1986, pp. 738]

Additionally, the Xbar and s chart is easier to interpret, and it checks the assumption that the population variances are equal rather than just assuming it as ANOVA does. If the data are time ordered, the control chart can discover trends and other patterns over time, which ANOVA fails to do.

Chapter Summary

- The within-subgroup estimate of sigma used for the control chart limits is sBar.

- Control limits for Xbar are narrower than control limits for X; some individual X values may fall outside of the Xbar limits.

112

- 3-sigma limit formulas (equal subgroup sizes):

$$UCL(s) = B_4 \bar{s}$$

$$LCL(s) = B_3 \bar{s}$$

$$UCL(\bar{X}) = \bar{\bar{X}} + A_3 \bar{s}$$

$$LCL(\bar{X}) = \bar{\bar{X}} - A_3 \bar{s}$$

- For time-ordered data, the Xbar and s chart with small subgroups may be used as a substitute for the I chart, but usually the I chart is a better choice.

- With the time-ordered Xbar and s chart for process improvement, all tests discussed are applicable.

- With the time-ordered Xbar and s chart to judge whether a state of control may be inferred, there should be 24 consecutive points within the 3-sigma limits. No other tests for lack of control are recommended.

- The Xbar and s chart and T-sigma limits are used for comparisons using *rational subgroups* where data within a subgroup are as much alike as possible. Examples: by gender, by physician, by shift, by hospital, by DRG, and so on.

- When using the Xbar and s chart with rational subgroups to judge whether a state of control may be inferred, there are as many subgroups as there are entities to compare.

- For charts with rational subgroups other than time-ordered subgroups, only the test for one or more points outside the T-sigma limits applies.

Problems

1.   For the following time-ordered data, make an Xbar and s chart. What do you conclude?

Day	1	2	3	4	5	6	7	8	9	10
Observation 1	27	23	27	24	29	22	41	23	27	23
Observation 2	27	23	23	19	24	23	45	24	23	23
Observation 3	22	24	29	24	25	21	39	20	29	24
Observation 4	24	25	22	23	24	21	40	24	23	24

2.   For the following time-ordered data, make an Xbar and s chart. What do you conclude?

Day	1	2	3	4	5	6	7	8	9	10
Observation 1	23	23	28	18	23	23	21	26	23	23
Observation 2	25	22	23	19	42	23	25	24	22	23
Observation 3	21	24	29	21	15	22	25	20	29	21
Observation 4	24	26	21	23	41	21	21	24	23	24

3. For the following data, make an Xbar and s chart by surgeon; by anesthesiologist. Use 3-sigma limits. (Use 1.5-sigma limits if using a computer.)

Surgery Time in Minutes	Surgeon	Anesthesiologist
115	A	C
140	A	D
90	A	C
170	A	D
95	A	C
165	B	D
75	B	C
95	B	D
105	B	C
180	B	D

4. For the following data, make an Xbar and s chart by DRG; by gender. Use 3-sigma limits. (Use 1.5-sigma limits if using a computer.)

Length of Stay in Days	DRG	Gender
2	A	M
1	A	F
4	A	M
1	A	F
2	A	M
4	B	F
6	B	M
2	B	F
5	B	M
3	B	F

5. Make an Xbar and s chart by surgeon. Use 3-sigma limits. (Use 1.5-sigma limits if using a computer.)
    Surgeon A: 130 95 105 105 55 115 95 155 135 140
    Surgeon B: 65 65 80 95 85 90 75 65 110 100

6. Make an Xbar and s chart by surgeon. Use 3-sigma limits. (Use 1.5-sigma limits if using a computer.) Is each surgeon in control over time?
    Surgeon A: 65 65 105 55 60 80 60 60 75 90 80 55 90 65 120 65 160 80 135 60 90 60 65 65 70
    Surgeon B: 145 70 100 70 125 80 135 165 65 125 75 70 105 90 65 75 100 65 140 80 85 120 70 65 70

7. One hundred turnaround times (in minutes) for CBCs from each of three shifts have been recorded, and their statistics are summarized below. Make an Xbar and s chart. Use 3-sigma limits. Use 2-sigma limits if using a computer. What do you conclude?

Shift	1	2	3
Xbar	35.0	37.9	36.8
s	4.72	5.0	2.51

8. Fifty turnaround times (in minutes) for CBCs from each of three shifts have been recorded, and their statistics are summarized below. Make an Xbar and s chart. Use 3-sigma limits. Use 2-sigma limits if using a computer. What do you conclude?

Shift	1	2	3
Xbar	27.0	35.2	36.4
s	4.3	5.0	2.5

Computer Supplement (Statit)

EXAMPLE 1

To illustrate the use of an Xbar and s chart for time-ordered data (for prediction), preliminary control limits will be set up as follows.

With data set: sem1.wrk
(*Note*: 3σ control limits will be set up for ongoing process control.)

> [Xs] button
>> Data variable: (click on ▶, Surgery_minutes, Done)
>> Subgroup: √ at Constant
>>> Size: type in *2*
>> Control limits
>>> Upper: Auto (the default)
>>> Center: Auto (the default)
>>> Lower: Auto (the default)
>> Chart title: Xbar/s Chart (the default)
>> Sub title: type in *on Surgery Times*

Note: 3σ control limits will be set up for ongoing process control. □

EXAMPLE 2

To illustrate the use of an Xbar and s chart for rational subgroups, reload sem2.wrk with information on Surgeon.

	Date	Surgery_minutes	Surgeon
1	04-Jan-2001	110	A
2	05-Jan-2001	120	A
3	09-Jan-2001	100	A
4	23-Jan-2001	90	A
5	25-Jan-2001	105	A
6	31-Jan-2001	130	B
7	14-Feb-2001	75	B
8	12-Mar-2001	95	B
9	04-May-2001	105	B
10	05-May-2001	103	B

To make Xbar and s charts to compare the two surgeons:
>[Xs] button
>>Data variable: (click on ▶, Surgery_minutes, Done)
>>Subgroup: √ at Grouped
>>>Variable: (click on ▶, Surgeon, Done)
>>Control limits
>>>Upper: Auto (the default)
>>>Center: Auto (the default)
>>>Lower: Auto (the default)
>>Chart title: Xbar/s Chart (the default)
>>Sub title: type in *on Surgery Times*
>>OK

Note: The number of sigmas used here should be 1.5σ limits (assuming there were more data from each surgeon), but the default of 3 will be used to check the hand calculations of Example 5.2. □

EXAMPLE 3

To illustrate the use of an Xbar and s chart with unequal subgroup sizes, add another variable (column) to sem2.wrk, so the data looks as follows:

	Date	Surgery_minutes	Surgeon	**Anesthesiologist**
1	04-Jan-2001	110	A	**D**
2	05-Jan-2001	120	A	**C**
3	09-Jan-2001	100	A	**C**
4	23-Jan-2001	90	A	**D**
5	25-Jan-2001	105	A	**C**
6	31-Jan-2001	130	B	**C**
7	14-Feb-2001	75	B	**D**
8	12-Mar-2001	95	B	**D**
9	04-May-2001	105	B	**C**
10	05-May-2001	103	B	**C**

Save the data as sem4.wrk
>File > Save Data File . . .
>>Save in: click on ▶, then 3 1/2 Floppy (A:) if desired
>>File name: type in *sem4*
>>OK

To make Xbar and s charts to compare the two anesthesiologists:
>Edit > Preferences > QC . . .
>>Number of sigmas: 1.5
>>OK
>[Xs] button
>>Data variable: (click on ▶, Surgery_minutes, Done)

Subgroup: √ at Grouped

          Variable: (click on ▶, Anesthesiologist, Done)

Control limits

    Upper: Auto (the default)

    Center: Auto (the default)

    Lower: Auto (the default)

Chart title: Xbar/s Chart (the default)

Sub title: type in *1.5 sigma limits*

OK

Note that the control limits vary for unequal-size subgroups on both the Xbar and the s charts and that the centerline varies on the s chart.

Remember that the number of sigmas for the next chart will now remain at 1.5 until it is changed again. ☐

Computer Supplement (Minitab)

EXAMPLE 1

To illustrate the use of an Xbar and s chart for time-ordered data (for prediction), preliminary control limits will be set up as follows.

With data set: sem1
(*Note*: 3σ control limits will be set up for ongoing process control.)

>   Stat > Control Charts > Xbar S . . .
>   Data are arranged as
>   Single column: (click on C2 Surgery_minutes, Select)
>   Subgroup size: (type in *2*)
>   OK ☐

EXAMPLE 2

To illustrate the use of an Xbar and s chart for rational subgroups, reload sem2 with information on Surgeon.

	Date	Surgery_minutes	Surgeon
1	1-4-01	110	1
2	1-5-01	120	1
3	1-9-10	100	1
4	1-23-01	90	1
5	1-25-01	105	1
6	1-31-01	130	2
7	2-14-01	75	2
8	3-12-01	95	2
9	5-4-01	105	2
10	5-5-01	103	2

To make Xbar and s charts to compare the two surgeons:
>   Stat > Control Charts > Xbar S . . .
>   Data are arranged as
>   Single column: (click on C2 Surgery_minutes, Select)
>   Subgroup size: (click on C3 Surgeon, Select))
>   OK

Note: The number of sigmas used here should be 1.5σ limits (assuming there were more data from each surgeon), but the default of 3 will be used to check the hand calculations of Example 5.2. ☐

EXAMPLE 3

To illustrate the use of an Xbar and s chart with unequal subgroup sizes, add another variable (column) to sem2 so the data looks as follows:

	Date	Surgery_minutes	Surgeon	Anesthesiologist
1	1-4-01	110	1	4
2	1-5-01	120	1	3
3	1-9-10	100	1	3
4	1-23-01	90	1	4
5	1-25-01	105	1	3
6	1-31-01	130	2	3
7	2-14-01	75	2	4
8	3-12-01	95	2	4
9	5-4-01	105	2	3
10	5-5-01	103	2	3

Save the data as sem4. Note that the data are not sorted by anesthesiologist. To sort by anesthesiologist use:
Manip > Sort . . .

Sort column(s): (type in: *C1-C4*)

Store sorted column(s) in: (type in *C1-C4*) (You could have typed in C5-C8 and typed in new column headings.)

Sort by column: (type in: *C4*)

OK

To make Xbar and s charts to compare the two anesthesiologists:
Stat > Control Charts > Xbar S . . .

Data are arranged as

Single column: (click on C2 Surgery_minutes, Select)

Subgroup size: (click on C4 Anesthesiologist, Select))

S-limits . . .

Sigma limit positions: (type in 1.5)

OK

Annotation (click on τ, Title . . .)

Title: (type in *1.5-sigma limits*)

OK

OK

Note that the control limits vary for unequal-size subgroups on both the Xbar and the s charts and that the centerline varies on the s chart. ☐

Minitab only gives the values for the centerlines and the control limits for the last subgroup. You can get the plotted points, centerlines, and control limits for up to only 16 subgroups to print in the session window by using the "brief" command as illustrated in Example 4.

You can get all the subgroup Xbar and *s* values by:
Stat > Basic Statistics > Display Descriptive Statistics . . .
        Variable: (Click on C2 Surgery_minutes, Select)
        √ at By variable: (check on C4 Anesthesiologist, Select)
        OK

EXAMPLE 4

To get a list of the plotted points, centerlines, and control limits, you must use the "brief" command and make the Xbar chart and the s chart separately.

To enable the command language, be in the session window (Window > Session). Then
        Editor > Enable Command Language
                at the MTB> prompt in the session window, type in *brief 6*
                hit Enter

Make the Xbar chart and the s chart separately.
        Stat > Control Charts > Xbar . . .
                Data are arranged as
                        Single column: (click on C2 Surgery_minutes, Select)
                        Subgroup size: (click on C4 Anesthesiologist, Select))
                S-limits . . .
                        Sigma limit positions: (type in 1.5)
                        OK
                Annotation (click on τ, Title . . .)
                        Title: (type in *1.5-sigma limits*)
                        OK
                OK

        Stat > Control Charts > S . . .
                Data are arranged as
                        Single column: (click on C2 Surgery_minutes, Select)
                        Subgroup size: (click on C4 Anesthesiologist, Select))
                  S-limits . . .
                        Sigma limit positions: (type in 1.5)
                        OK
                Annotation (click on τ, Title . . .)
                        Title: (type in *1.5-sigma limits*)
                        OK
                OK

Scroll through the session window to view the list of the plotted values, centerlines, and control limits. □

Note: When using standard given for an Xbar and s chart, σ may be estimated by $\dfrac{\sqrt{n}(UCL - \overline{\overline{X}})}{3}$.

Case Study 5.1   Turnaround Times for CBCs

The concept for this case study was developed with the assistance of Dr. Ray Carey of R. G. Carey and Associates, Park Ridge, IL.

Background

There have been complaints about the turnaround times (TATs) for complete blood cell counts (CBCs). The quality assurance team decided to look at available data on TATs to assess the situation. Presently, the only historical data available are average TATs for each of 3 shifts. The team decided to study the data from the most recent four weeks (28 days).

The Data

(Data are in files noted.) The average TAT values for each of three shifts for the most recent four calendar weeks, rounded to the nearest minute, are shown in Table 1. The quality of the data is impaired because the number of measurements that have been averaged to obtain each tabular entry is not known. However, it is presently necessary to work with this imperfect data while collecting the individual measurements for future study.

Case Study 5.1 Table 1 Average TAT by Shift for CBCs

Day	Average TAT			Date	Day
	Shift 1	Shift 2	Shift 3		
1	56	43	59	4/2	Sunday
2	53	55	50	4/3	Monday
3	112	63	56	4/4	Tuesday
4	56	66	60	4/5	Wednesday
5	73	45	54	4/6	Thursday
6	53	54	66	4/7	Friday
7	135	56	56	4/8	Saturday

8	124	53	75	4/9	Sunday
9	102	66	45	4/10	Monday
10	121	76	72	4/11	Tuesday
11	62	46	51	4/12	Wednesday
12	50	48	48	4/13	Thursday
13	58	51	39	4/14	Friday
14	65	74	57	4/15	Saturday
15	59	59	37	4/16	Sunday
16	63	53	69	4/17	Monday
17	70	51	45	4/18	Tuesday
18	60	54	36	4/19	Wednesday
19	52	39	39	4/20	Thursday
20	52	37	31	4/21	Friday
21	62	62	48	4/22	Saturday
22	108	44	46	4/23	Sunday
23	56	52	47	4/24	Monday
24	46	42	58	4/25	Tuesday
25	44	41	43	4/26	Wednesday
26	51	67	34	4/27	Thursday
27	44	40	44	4/28	Friday
28	42	70	38	4/29	Saturday
Xbar	68.9	53.8	50.1		
s	27.1	10.9	11.6		

Analysis, Results, and Interpretation

As a first step in analysis, it is always wise to study the tabular data carefully to see if there are indications to suggest nonrandom influence. An easy way to do this is to look for any pattern in the extreme values that could not reasonably have arisen by random chance. The reader is encouraged to study the data in Table 1 at this time. What do *you* see?

Sometimes it is useful to circle the highest values to see if any clear pattern emerges. In Table 1, this is hardly necessary; there are only six three-digit numbers in the 84 tabular entries, and they all occur on the first shift. It should be clear that this distribution of the extreme values is a nonrandom pattern. If you want computational evidence, note that the highest of the 84 values has to occur in some column—and it happens to be shift 1. The probability of the second highest occurring in the same shift by random chance is 27/83 since 27 of the remaining 83 values are from shift 1. The probability of the six highest values all occurring in shift 1 by random chance is $(27/83)(26/82)(25/81)(24/80)(23/79) = 0.00278$. Patterns in the extreme values suggest nonrandom influence. The TATs are stratified by shift, but there is more information to be found in the data. Continuing as if there had been no perusal of the data, Figure 1 is a run chart on the average TAT data in time order. Note that there is no clear indication *from this chart* of the reason for the frequent spikes that occur.

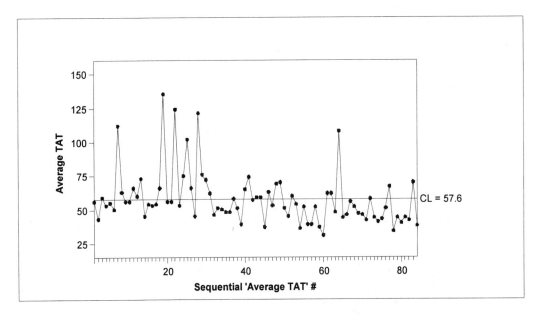

Case Study 5.1 Figure 1 Run Chart for Average CBC TAT, All Three Shifts in Time Order (Data Are in File ZTATC)

The *interrupted* run chart on the average TAT values, Figure 2, is an improvement over Figure 1. In Figure 2 the line connecting the plotted points is interrupted between shift 3 on one day and shift 1 on the next day. Now it is clear that each of the spikes occurred on the first shift. This is clearly not a random pattern, as was noted in the study of the extreme values in the tabular data. A special cause of variation has been found, which should be eliminated as soon as possible. In the meantime, this special cause must be taken into account in the analysis: all three shifts may not be pooled together for any analyses.

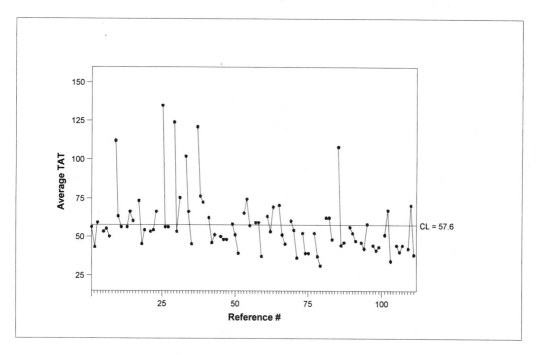

Case Study 5.1 Figure 2 Interrupted Run Chart for Average CBC TAT, All Three Shifts in Time Order, Interruptions between Days (Data Are in File ZTAT1)

It still is necessary to study each shift by itself in time order for indications that suggest nonrandom influence. I charts are now made on each shift separately, recognizing that the question of each data set being near-normal still needs to be addressed. Figures 3, 4, and 5 are I charts with no standard given on shifts 1, 2, and 3, respectively.

It may be inferred from Figure 3 that shift 1 had a process improvement after day 10. (This may be verified in problem 3 at the end of this case study.) It follows that the control limits, having been calculated across a discontinuity in the process, are not valid. Since there were two different processes within the shift 1 data set, there is nothing to be gained by pursuing the normality question here. In Figure 4 there are no indications for shift 2 that suggest nonrandom influence. The data have a near-normal distribution, as the reader may verify, so the control limits are valid. Figure 5 suggests that shift 3 may have had a process improvement after day 16. If this were to be verified by an Xbar and s chart, the control limits in Figure 5 should be updated after the process improvement. (See the problem relating to this at the end of this case study.)

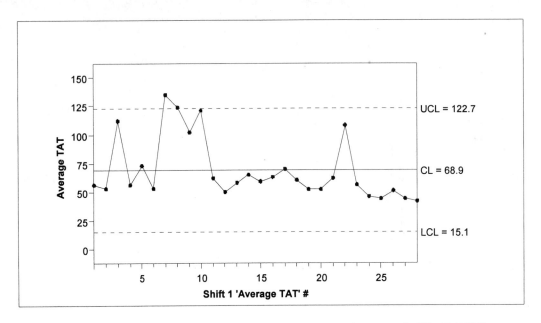

Case Study 5.1 Figure 3 I Chart for Shift 1 Average CBC TAT (Data Are in File ZTATC)

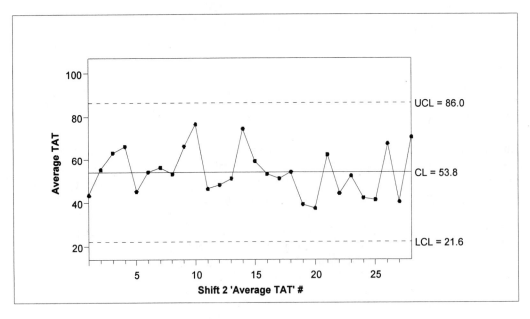

Case Study 5.1 Figure 4 I Chart for Shift 2 Average CBC TAT (Data Are in File ZTATC)

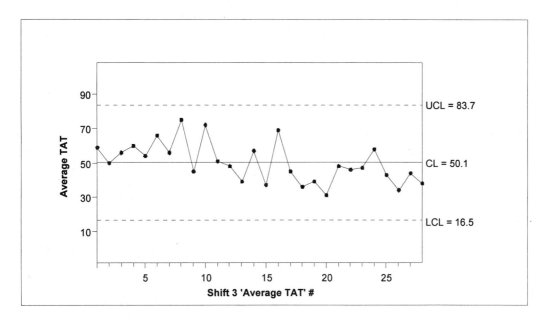

Case Study 5.1 Figure 5 I Chart for Shift 3 Average CBC TAT (Data Are in File ZTATC)

Figure 6 is an Xbar and s chart using 2-sigma limits to compare the three shifts for the 28-day period. This chart substantiates what has already been learned, that shift 1 has significantly higher average TAT values than the other two shifts. But now this has to be interpreted in light of the evidence that shift 1 can do as well as the others. The high TAT measures for shift 1 are apparently not built into the system since it has been shown that improvement can be made.

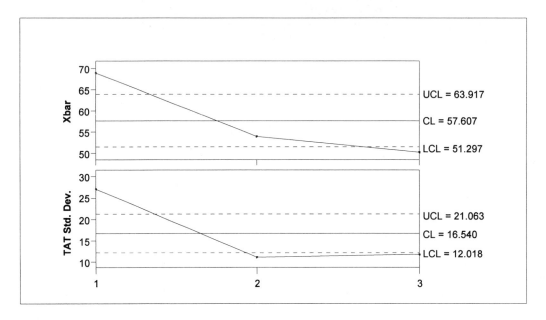

Case Study 5.1 Figure 6 Xbar and s Chart on Average CBC TAT with $n = 28$, Subgrouped by Shift, 2-Sigma Limits (Data Are in File ZTATC)

Lessons Learned

Besides analyzing the data by rational subgroups, each subgroup of time-ordered data should be studied separately in time order. Here an I chart was used for each shift in turn. In the future, the individual TAT values should be recorded, instead of just the shift averages of these values.

Managerial Implications

Whatever improvements were made in shifts 1 and 3 should be investigated and perhaps incorporated as standard procedure.

When actively working to improve a process with the use of an I chart, a separate I chart must be kept for ongoing process control on each of the three shifts. However, when using a run chart for process maintenance there are two possibilities:

1. Keep a separate chart for each shift.
2. Keep a single chart for the three shifts, using interruptions between the shifts and a different symbol (or other annotation) for each shift.

Three separate run charts may have the advantage of possibly catching the deterioration of a single shift more quickly. However, the three separate run charts have the disadvantage of chart proliferation, plus the fact that a process change that affects all three shifts may go unnoticed for a longer period of time.

Experience has shown that the single chart is generally better due to simplicity for process maintenance, but there is room here for personal preference.

Process capability could not be established since the data available were average TATs rather than individual TATs. This is reason enough to want the individual TAT values recorded.

Problems

1.   Looking at Figures 3 through 5 above, it seems likely that the special-cause variation between the three shifts may have disappeared in the final two-week period. Using an approach similar to that of Figure 6, what can be concluded?

2.   A common approach (unfortunately) to the data in Table 1 would be to make a time-ordered control chart with $k = 28$ subgroups (days) of size $n = 3$ (shifts). However, with such a chart, instead of having homogeneous subgroups, there would be within-subgroup stratification that would result in inflated control limits. Make such a "wrong" Xbar and s chart and compare the results with the "right" analyses made in this case study.

3.   Using an Xbar and s chart, verify that improvement was made by shift 1 after day 10.

4.   Does an Xbar and s chart indicate that improvement was made by shift 3 after day 16?

5.   Verify that the shift 2 data are near-normal.

Chapter 6 Process Capability

Introduction to Process Capability Estimates

With variables data, the *capability* of a process may be defined in terms of an upper and lower limit to the population distribution such that no more than about one observation in 1,000 is expected in each tail. These are the limits that the process is said to be "capable" of meeting. This definition of process capability is consistent with the 3-sigma limits for a normal distribution (called the "natural limits"), where about 0.135% is expected in each tail. If the value of the individual readings cannot physically take on values less than a given value (X min), then the lower process capability limit may, in practice, be assigned the value X min.

It may be important to establish the process capability because there may be no way to tell if the process needs to be improved unless its present capability is known. Being in control only establishes the stability of the process; it does not imply that the process is satisfactory. Also, if changes are made in the process, there is no way to determine whether these changes are an improvement unless the process capability is known before and after the change.

Three methods of estimating process capability are considered here. The first two methods assume a normally distributed process and so are called *normal estimates of process capability*.

1. If the data were normally distributed, the 3-sigma control limits of an I chart would provide one estimate of the natural limits, which may be used as the first normal estimate of process capability.
$$UNL = UCL(X)$$
$$LNL = LCL(X)$$

2. Again assuming a normal distribution, Xbar plus and minus three times the standard deviation (s) of all of the observations provides another estimate of the natural limits, which may be used as the second normal estimate of process capability.
$$UNL = Xbar + 3s$$
$$LNL = Xbar - 3s$$

3. The third estimate of process capability is obtained by finding the 0.00135 and 0.99865 quantiles graphically from the probability plot.

If the value of the individual readings cannot physically take on values less than a given value (X min), then the lower process capability limit may, in practice, be assigned the value X min. It will be shown in the following that the best estimate of process capability comes from the probability plot. It works well regardless of the shape of the distribution. This is the method recommended here.

EXAMPLE 6.1

Example 2.12 gave a set of 10 lap chole surgery times in minutes with the probability plot shown in Example 2.13 as Figure 2.8. Process capability is estimated here by the three methods for these data. Such a small amount of data is useful only for tutorial purposes to illustrate the calculations. The estimated process capabilities have been rounded to the nearest minute.

1. The first normal estimate assumes a normal distribution and uses the limits from the I chart (Figure 4.1, repeated here as Figure 6.1). The I chart control limits are the natural limits of the process.

    $$UNL = UCL(X) = 153 \text{ minutes}$$
    $$LNL = LCL(X) = 54 \text{ minutes}$$

    First normal process capability estimate = [54, 153] minutes.

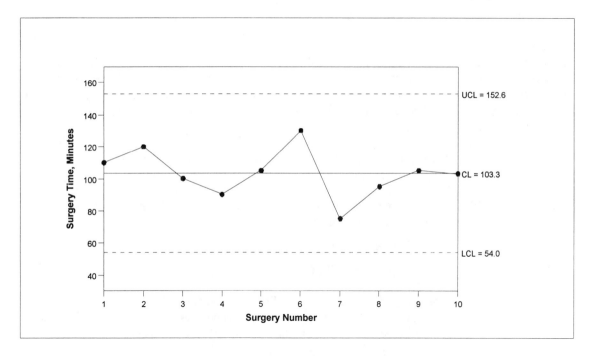

Figure 6.1 I Chart Used to Estimate Process Capability, 10 Observations

2. The second normal estimate assumes a normal distribution and uses Xbar $\pm 3s$.

    From Example 2.8, Xbar = 103.3 and s = 15.28. This yields estimates of the natural limits of

    $$UNL = Xbar + 3s = 103.3 + 3(15.28) = 103.3 + 45.84 = 149$$
    $$LNL = Xbar - 3s = 103.3 - 3(15.28) = 103.3 - 45.84 = 57$$

    Second normal process capability estimate = [57, 149] minutes.

3.　　Best estimate using the probability plot: The probability plot in Figure 2.8 has been modified in Figure 6.2 with the graphical work needed to obtain the process capability. The probability plot is the best-fit smooth curve. If that smooth curve happens to be a straight line, it should be considered fortuitous. There is so little data in this example that an "eyeball" best-fit straight line fits as well as anything, as illustrated in Figure 6.2. (The subject of an "eyeball" best fit is covered in more detail below.)

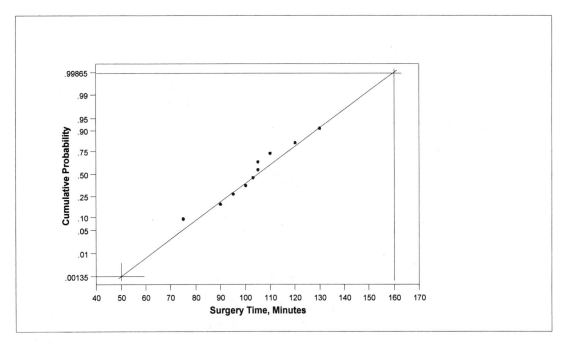

Figure 6.2 Probability Plot Used to Estimate Process Capability, 10 Observations

A horizontal line is drawn from the 0.99865 mark on the y-axis until it intersects the best-fit line (or best-fit curve). From that point of intersection, a vertical line is dropped until it intersects the x-axis, here at approximately 160. This value of 160 is the 0.99865 quantile (approximately the 99.9th percentile), the value that will be exceeded only about once in a thousand observations. Similarly, a horizontal line is drawn from the 0.00135 mark on the y-axis until it intersects the best-fit line. From that point of intersection, a vertical line is dropped until it intersects the x-axis, here at approximately 50. The point where this intersection occurs is then the 0.00135 quantile (approximately the 0.1th percentile), that is, the value below which only about one observation in a thousand will fall.

Best process capability estimate = [50, 160] minutes.

It should be emphasized that this example with a sample of only 10 and a straight line for best fit is solely for an initial illustration of the computational methods. More realistic applications are given in the following examples and in the case studies. □

133

It is suggested that the reader study again the section in Chapter 2 titled "Making the Probability Plot" before proceeding to the next section.

Process Capability from a Straight-Line Probability Plot

For a normal distribution, the probability plot will tend toward a straight line. The first step is to draw the "eyeball" best-fit straight line. Shapiro [1990, p.11 ff] gives the following comments on the fitting of a straight line for a probability plot:

> It is often helpful . . . to fit by eye a "best line" through the plotted points. . . . the following should be kept in mind:
>
> i. The observed values are random variables and hence will never lie perfectly on a straight line.
> ii. The ordered values are not independent since they have been ranked . . . Therefore, if one point is above the line there is a good chance that the next one will also be above the line. . . . There could be significant runs . . . above and below the line even though the data came from a normal population.
> iii. The variances of the extreme points (the largest and smallest) are much higher than the points in the middle of the plot, and hence a greater discrepancy from a straight line should be allowed at the two extremes. When drawing a line through the data the center points should be given more weight than the extremes.

Note that the task of finding the eyeball best-fit line cannot be assigned to the computer because there is no computer algorithm for the best-fit criteria used here. Further note that with regard to item iii above, only the eyeball best-fit straight line on the probability plot can give the proper weighting of points at the center and extremes. It follows that the probability plot results will be a better estimate than either of the two normal estimates. However, if the process generating the data really is in control and really is normally distributed, all three methods will yield similar results as illustrated by the following example.

The three methods of estimating process capability will be illustrated for this data set in Example 6.2. The advantage of using synthesized data is that the results may be compared with truth that was known beforehand, a privilege not available with real process data.

EXAMPLE 6.2

The first of the three synthesized data sets considered here were first discussed in Chapter 2. It consists of 200 observations from a synthesized normal distribution with the histogram and probability plots shown in Figures 6.3 and 6.4 repeated from Figures 2.5 and 2.9.

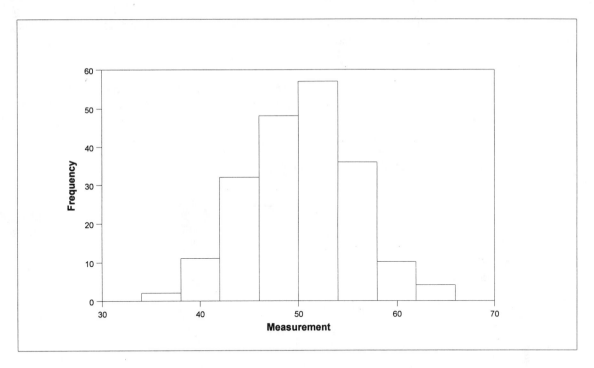

Figure 6.3 Histogram of Synthesized Normal Distribution, 200 Observations

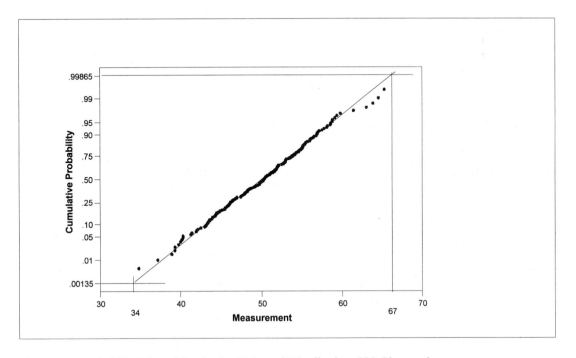

Figure 6.4 Probability Plot of Synthesized Normal Distribution, 200 Observations

The data in the histogram of Figure 6.3 appear to be approximately normally distributed, and this is verified by the probability plot in Figure 6.4. Using the three methods of estimating the natural limits, we get the following estimates:

1. Using the control limits from the I chart, Figure 6.5, gives the first normal process capability estimate. Note that all points fall within the control limits, which is consistent with the synthesized normal distribution.

    UNL = UCL(X) = 66.22
    LNL = LCL(X) = 34.12
    First normal process capability estimate = [34, 66].

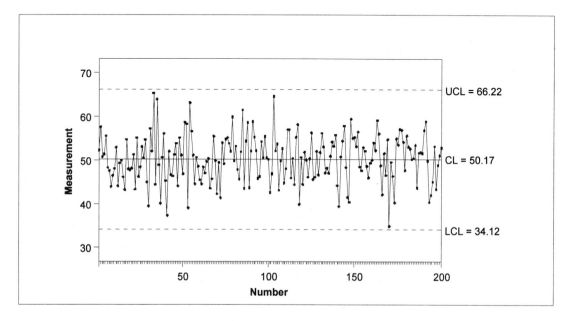

Figure 6.5 I Chart of Synthesized Normal Distribution, 200 Observations

2. Given Xbar = 50.17 and s = 5.436, the second process capability estimate is
    UNL = Xbar + 3s = 50.17 + 3(5.436) = 50.17 + 16.31 = 66.48
    LNL = Xbar - 3s = 50.17 - 3(5.436) = 50.17 - 16.31 = 33.86
    Second normal process capability estimate = [34, 66].

3. The best-fit line through the plotted points on the probability plot is shown in Figure 6.4. Note the departure of the extreme points from that line, consistent with Shapiro's item iii. Also note near the upper end of the line that when one point is low, the next one also tends to be low, Shapiro's item ii. The 0.00135 and 0.99865 quantiles in Figure 6.4 yield
    Best process capability estimate = [34, 67].

136

The three estimates found above are shown in the first row of Table 6.1 later in this chapter. Note that with 200 observations from a simulated normal distribution, the two normal estimates give good agreement with the best estimate from the probability plot. □

Shewhart [1939] recommended no less than 100 observations for estimating the process capability from synthesized normal numbers. His recommendation seems compatible with the estimation process using 200 observations in Example 6.2. The difficulty in estimating process capability with only 100 or 200 points can be understood by noting the need to extrapolate the best-fit curve down to the 0.00135 quartile and up to the 0.99865 quartile; the smaller the quantity of data, the greater the length of extrapolation required and the less certainty you can have in the results. Since you know in advance with synthesized normal data that the best fit is a straight line, there is much greater confidence in the extrapolation.

Process Capability from a Smooth Curve Probability Plot

In the general case, the probability plot will not be a straight line, but a smooth curve that is the eyeball best fit to the plotted data points. Shapiro's comments on the fitting of a straight line for a probability plot are equally helpful for fitting a smooth curve. Again, this task cannot be assigned to the computer because there is no computer algorithm for the best-fit criteria.

Example 6.3 is the first example using a curved probability plot. The data set is the second of the three synthesized sets of size $n = 200$ from Chapter 2 and is moderately skewed to the right.

EXAMPLE 6.3

The data in the histogram (Figure 6.6) are not normally distributed. The data are synthesized to be moderately skewed to the right, as can be seen on the histogram and probability plot, Figures 2.6 and 2.10, repeated here as Figures 6.6 and 6.7. Using the three methods of estimating the natural limits yields the following estimates:

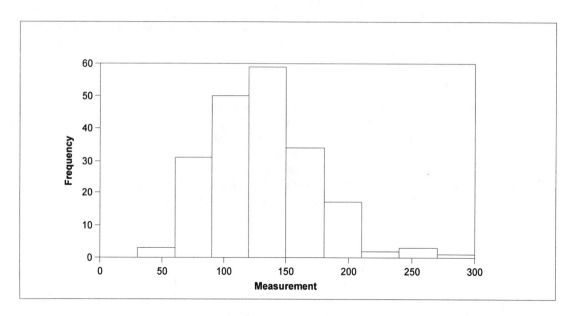

Figure 6.6 Histogram of Synthesized Moderately Skewed Distribution, 200 Observations

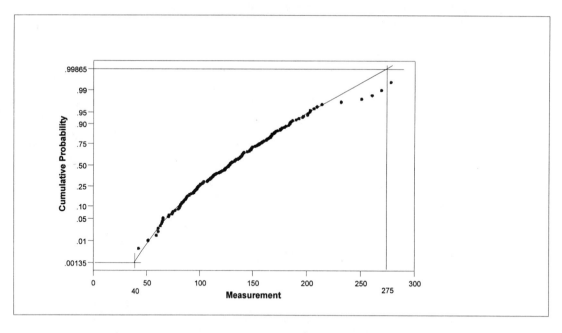

Figure 6.7 Probability Plot of Synthesized Moderately Skewed Distribution, 200 Observations

1.   On the I chart (Figure 6.8) there are three points clearly above the 3-sigma UCL and no points even close to the LCL. This is consistent with the moderate skew to the right. In classifying this data set as near-normal and satisfactory for the use of the control chart, it should be recognized

that the control chart use here is marginal. The three outages in 200 points demonstrate a false alarm risk of 0.015, which is acceptable. To repeat a quote from Chapter 4, the ANSI Standard Z1.3 [1958, 1975, p. 18] states:

> it is usually not safe to conclude that a state of control exists unless the plotted points for at least 25 successive subgroups fall within the 3-sigma control limits. In addition, if not more than 1 out of 35 successive points, or not more than 2 out of 100, fall outside the 3-sigma control limits, a state of control may ordinarily be assumed to exist.

If the objective were to determine if a state of control exists, using Table 4.1, 4-sigma limits could be used (Figure 6.9) and a state of control would be inferred. The decision to use a control chart on this data is based on the fact that it will still work reasonably well and the data can be considered near-normal. The alternative, transforming the data (see Chapter 9), has its own problems.

Using the 3-sigma control limits from Figure 6.8:
UNL = UCL(X) = 251.27
LNL = LCL(X) = 10.15
First normal process capability estimate = [10, 251].

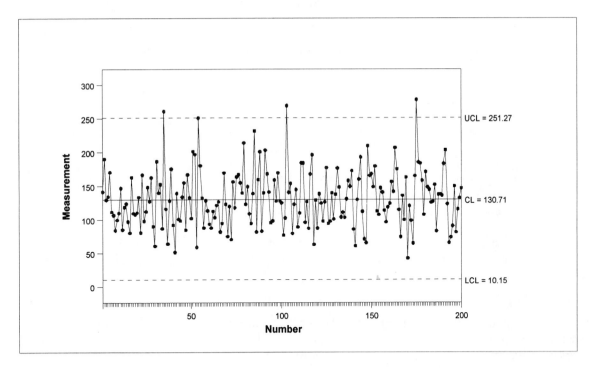

Figure 6.8 I Chart (3-Sigma) of Synthesized Moderately Skewed Distribution, 200 Observations

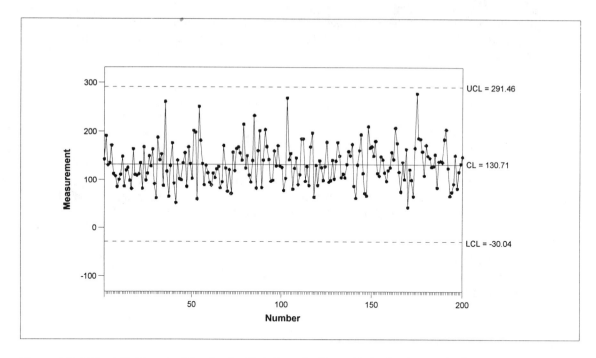

Figure 6.9 I Chart (4-Sigma) of Synthesized Moderately Skewed Distribution, 200 Observations

2. Given Xbar = 130.71 and s = 41.978
    UNL = Xbar + 3s = 130.71 + 3(41.978) = 130.71 + 125.93 = 256.64
    UNL = Xbar - 3s = 130.71 - 3(41.978) = 130.71 - 125.93 = 4.78
    Second normal process capability estimate = [5, 257].

3. Drawing a best-fit line through the plotted points on the probability plot as shown in Figure 6.7, the 0.00135 and 0.99865 quantiles yield:
    Best process capability estimate = [40, 275].

The three estimates found above are shown in the second row of Table 6.1 later in this chapter. Note that with 200 observations from a simulated distribution that is moderately skewed, the two normal estimates are of little value. □

In Chapter 2 a "near-normal" distribution was defined as one where the departures from normality were so sufficiently small that the control chart would still satisfactorily discriminate between common-cause and special-cause variation. From Example 6.3 it can be observed that the estimate of process capability is *not* forgiving with respect to departures from normality; even moderate departures from normality have large effects.

It is instructive to note the substantial degree of uncertainty that exists in Figure 6.7, even knowing that the extreme points of the plot are not to be trusted. Two hundred points here are clearly not enough to estimate process capability with reasonable confidence. Shewhart [1939] recommended no less than 1,000 observations with particular care to have the observations gathered under all possible circumstances (all physicians, all severities of patient illness, all shifts, etc.) to estimate the capability of real processes. In practice, process capability estimates are usually made with far less than the recommended 1,000 observations. Sometimes this is because it is not economically feasible to get more data. Sometimes it is because it is not possible to go back far enough in time before finding a process change that would disallow earlier data. Under these circumstances you must be sure to obtain data from the current process under all possible circumstances and then live with the uncertainty of the prediction. Regardless of how much or how little data is available, the best estimate of process capability will come from the probability plot.

Example 6.4 uses the last of the three simulated data sets of size $n = 200$ in Chapter 2.

EXAMPLE 6.4

This process is severely skewed to the right, as can be seen on the histogram and probability plot, Figures 2.7 and 2.11, repeated here as Figures 6.10 and 6.11. The three estimates of process capability are shown below.

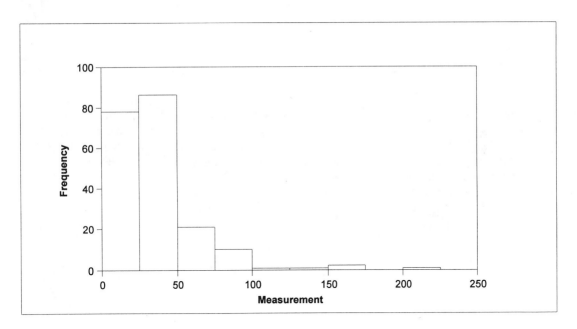

Figure 6.10 Histogram of Synthesized Severely Skewed Distribution, 200 Observations

141

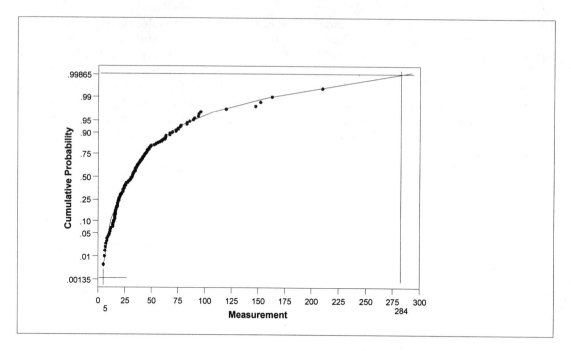

Figure 6.11 Probability Plot of Synthesized Severely Skewed Distribution, 200 Observations

1. On the I chart (Figure 6.12) there are five points above the 3-sigma UCL and no points more than halfway from the centerline to the LCL. Even if 4-sigma limits are used (per Table 4.1), the I chart in Figure 6.13 shows lack of control. This is all consistent with the severe skew to the right. Here the data are too badly skewed for the use of a control chart without transforming the data. (See Chapter 9.) Using the 3-sigma control limits from Figure 6.12

        UNL = UCL(X) = 104.62

        LNL = LCL(X) = -29.73

First normal process capability estimate = [-30, 105].

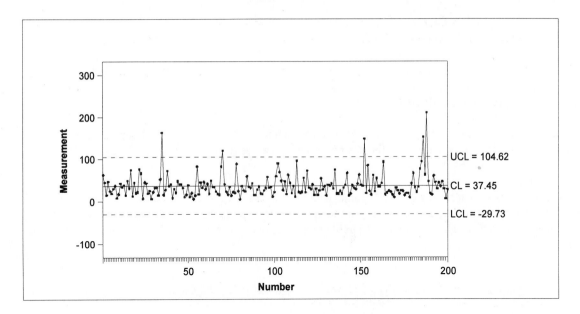

Figure 6.12 I Chart (3-Sigma) of Synthesized Severely Skewed Distribution, 200 Observations

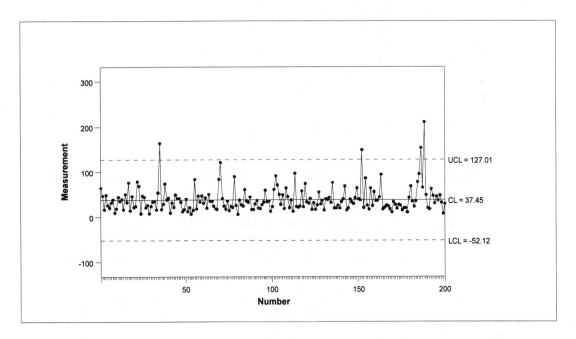

Figure 6.13 I Chart (4-Sigma) of Synthesized Severely Skewed Distribution, 200 Observations

2.   Given Xbar = 37.45 and s = 27.887,
              UNL = Xbar + $3s$ = 37.45 + 3(27.887) = 37.45 + 83.66 = 121.11

LNL = Xbar - $3s$ = 37.45 - 3(27.887) = 37.45 - 83.66 = -46.21
Second normal process capability estimate = [-46, 121].

3.   Drawing a best-fit line through the plotted points on the probability plot as shown in Figure 6.11, the 0.00135 and 0.99865 quantiles yield:
Best process capability estimate = [5, 284].

Compare these results in Table 6.1 with those in the preceding two examples. Note that with this severely skewed distribution the two normal estimates are so far in error as to be worthless. □

Note in Figure 6.11 that having only 200 points with this much skew leaves far more uncertainty in the extrapolation than would be desired. The extrapolation becomes particularly troublesome as the distribution becomes more skewed. This is why it is desirable to have as much data as possible (gathered over the same circumstances).

Table 6.1 Summary of Results from Three Synthesized Distributions, Each with $n = 200$

Synthesized Distribution	Normal Estimates		Probability Plot
	Method 1	Method 2	
Normal	[34, 66]	[34, 66]	[34, 67]
Moderately skewed to the right	[10, 251]	[5, 257]	[40, 275]
Severely skewed to the right	[-30,105]	[-46,121]	[5,284]

Population Distributions and Their Probability Plots

Figure 6.14 illustrates a perfect normal distribution and its best-fit smooth curve, which in this special case is a straight line. Straight lines for a best fit may occur occasionally in health care but will not be the general rule.

Figure 6.15 illustrates a process that is skewed to the right, yielding a smooth curve that is concave downward. Such distributions may occur when there is a lower bound to the data. Time intervals, for example, have a lower bound of zero but no upper bound. Many healthcare processes will be similar to Figure 6.15.

Figure 6.16 shows a process where the distribution is skewed to the left, yielding a smooth curve that is concave upward. Such distributions are not often seen in health care but may occur when there is an upper bound that the data cannot exceed.

Figure 6.17 illustrates the case of a bimodal distribution. This may occur when two different distributions have been improperly pooled, for example, pooling together variables data from two hospitals that differ significantly.

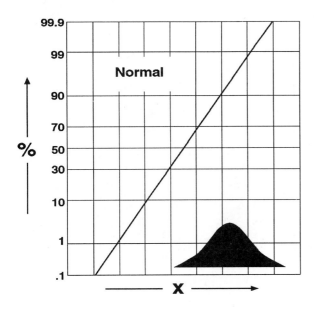

Figure 6.14 Probability Plot: Normal Distribution (Source: From *Quantitative Methods for Quality and Productivity Improvement*, by M. Hart and R. Hart, p. 229. Copyright © 1989 ASQC Quality Press.)

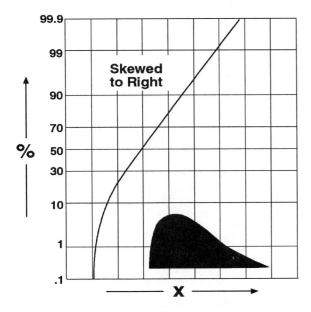

Figure 6.15 Probability Plot: Skewed to the Right (Source: From *Quantitative Methods for Quality and Productivity Improvement*, by M. Hart and R. Hart, p. 229. Copyright © 1989 ASQC Quality Press.)

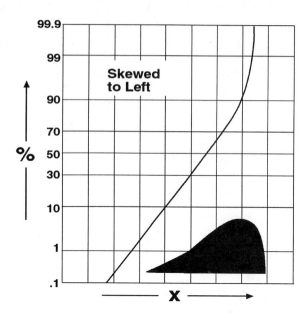

Figure 6.16 Probability Plot: Skewed to the Left (Source: From *Quantitative Methods for Quality and Productivity Improvement*, by M. Hart and R. Hart, p. 229. Copyright © 1989 ASQC Quality Press.)

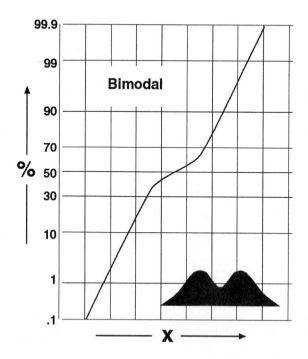

Figure 6.17 Probability Plot: Bimodal (Source: From *Quantitative Methods for Quality and Productivity Improvement*, by M. Hart and R. Hart, p. 230. Copyright © 1989 ASQC Quality Press.)

Decreasing Variability

Improving quality is related to decreasing variability [Taguchi, 1981]. A decrease in variability implies more consistency. As an illustration, consider the case of scheduling an operating room for a given procedure. Surgeon A averages 150 minutes; surgeon B averages 140 minutes. So schedule less time for surgeon B? But here is the rest of the story. Both have times that are skewed to the right. Surgeon A's process capability is [130, 200]; surgeon B's process capability is [60, 250]. So the answer is no, allow more time for surgeon B. Surgeon A is actually easier to schedule due to less variability.

Chapter Summary

Process capability is defined in terms of an upper and lower limit to the population distribution such that no more than about one observation in 1,000 is expected in each tail.

- There are three methods of estimation. With a normal distribution assumption:
  1.   The control limits on the I chart
  2. $Xbar \pm 3s$
  With no distribution assumption (recommended here):
  3.   the 0.00135 and the 0.99865 quantiles on the probability plot.

- If the process is *really* in control and normally distributed, all three estimates should give similar results. (*Note*: Being in control only establishes the stability of the process. It does not imply that the process capability is satisfactory.)

- Improving quality is related to decreasing variability.

Problems

1.   Given the following times (in hours) for transcription, calculate the process capability using the three methods. What do you conclude?
14 25 26 34 16 46 27 13 22 21 34 13 1 4 8 21 34 19 9 11 27 8 30 23 28

2.   Given the following times (in hours) for transcription, calculate the process capability using the three methods. What do you conclude?
46 27 15 22 21 14 25 26 34 36 21 32 19 19 17 34 15 29 24 20 27 20 30 23 29

3.   For the following 39 time-ordered surgery times (in minutes) for lap choles by surgeon E, estimate process capability by the three methods. Which is the most appropriate?
250 100 75 75 70 155 65 60 75 70 88 90 115 90 75 75 120 125 85 80 105 95 80 85 150 100 85 75 80 80 160 85 110 100 120 210 85 90 270

4.   For the following 35 time-ordered surgery times (in minutes) for lap choles by surgeon F, estimate process capability by the three methods. Which is the most appropriate?
65 65 105 55 60 80 60 60 75 90 80 55 90 65 120 65 168 80 135 60 90 60 85 62 70 50 95 120 70 60 70 60 75 90 52

Computer Supplement (Statit)

There are three methods presented for estimating process capability. The first two (control limits on the I chart and Xbar $\pm$ 3s) have normality assumptions. The third method (probability plot) makes no assumption about the shape of the distribution and is, hence, preferred. These three methods are revisited here.

EXAMPLE 1

Method 1: Individual Chart:
With data set: sem1.wrk
     [I] button
               Data Variable: (click on ►, Surgery_minutes, Done)
               X Labels variable: (click on ►, Date, Done)
               X-Axis label frequency: (type in *1*)
               Chart title: (type in *Surgery Times*), Sub title: (type in *minutes*)
               OK ☐

EXAMPLE 2

Method 2: Calculating Average and Standard Deviation:
With data set: sem1.wrk
     Statistics > Descriptive . . .
               Variable: (click on ►, Surgery_minutes, Done)
               Output Format: √ at Table
               Statistics: leave mean and standard deviation checked, delete checks at other statistics
               OK ☐

EXAMPLE 3

Method 3: Probability Plot:
With data set: sem1.wrk
     Graphs > Probability . . .
               Variable: (click on ►, Surgery_minutes, Done)
               delete √ at Plot a theoretical quantiles line
               delete √ at Perform test for normality
               Axis endpoints:
                     √ at .00135 - .99865
               OK ☐

Computer Supplement (Minitab)

There are three methods presented for estimating process capability. The first two (control limits on the I chart and Xbar $\pm$ 3s) have normality assumptions. The third method (probability plot) makes no assumption about the shape of the distribution and is, hence, preferred. These three methods are revisited here.

EXAMPLE 1

Method 1: Individual Chart:
With data set: sem1
        Stat > Control Charts > Individuals . . .
                Variable: (click on C2 Surgery_minutes, Select)
                OK ☐

EXAMPLE 2

Method 2: Calculating Average and Standard Deviation;
With data set: sem1
        Stat > Basic Statistics > Display Descriptive Statistics . . .
                Variable: (click on C2 Surgery_minutes in left column, Select)
                OK ☐

EXAMPLE 3

Method 3: Probability Plot:
With data set: sem1
        Graph > Probability Plot . . .
                Variables: (click on C2 Surgery_minutes, Select)
                Distribution: Normal
                Options . . .
                        delete √ at Include confidence intervals in plot
                        OK
            OK ☐

Case Study 6.1 Surgery Times

Background

Scheduling of the operating room has been a problem, and it has been determined that a contributing factor is the large variation in the surgery time for laparoscopic cholecystectomies (lap choles). Several factors of this problem were studied. Total time associated with the surgery was broken into setup time, pre-surgery time, and surgery time. The one considered here is the surgery time.

The Data

(Data are in file ZSURGERY.) For each lap chole procedure, the surgeon and the surgery time (in minutes) was recorded along with additional information. Time is measured to the nearest minute, so the data are variables data. A total of 351 consecutive lap chole procedures were recorded over a one-year period. Nine surgeons were involved. The least number of lap chole procedures done by a surgeon during this time period was 18; the most was 66. The data for surgeon A are shown in Table 1.

Case Study 6.1 Table 1 Surgery Times for Surgeon A

Surgery Number	Surgery Time (in minutes)	Surgery Number	Surgery Time (in minutes)
1	65	31	80
2	75	32	115
3	75	33	75
4	60	34	95
5	165	35	130
6	75	36	125
7	85	37	105
8	85	38	70
9	80	39	72
10	95	40	95
11	65	41	90
12	65	42	120
13	85	43	75
14	68	44	90
15	190	45	85
16	120	46	90
17	105	47	80
18	115	48	115
19	58	49	80
20	70	50	65
21	80	51	70
22	75	52	185
23	65	53	75
24	75	54	90
25	90	55	120
26	75	56	130
27	65	57	65
28	50	58	65
29	80	59	115
30	75	60	65

Analysis, Results, and Interpretation

First checking the assumption of near-normality for surgeon A, the probability plot (Figure 1) and the histogram (Figure 2) show that the data are severely skewed to the right. Results for the six surgeons with $n > 24$ are similar. It appears that the surgery times are by nature severely skewed. There is some lower bound (about 40 minutes) on the times, but there is no upper limit. Complications can add a significant amount of time.

The effect of this severe departure from normality renders the limits on the I chart for surgeon A (Figure 3) invalid. The control chart provides no clue as to whether the three points above the control limit are due to the skewed distribution or to the process being unstable over time. That determination will have to await the subject of transformations (Chapter 9 and Case Study 9.1).

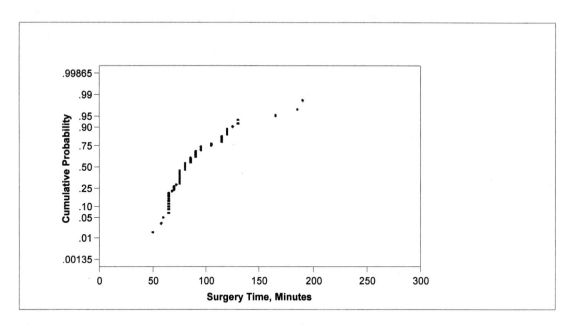

Case Study 6.1 Figure 1 Probability Plot of Lap Chole Surgery Times for Surgeon A

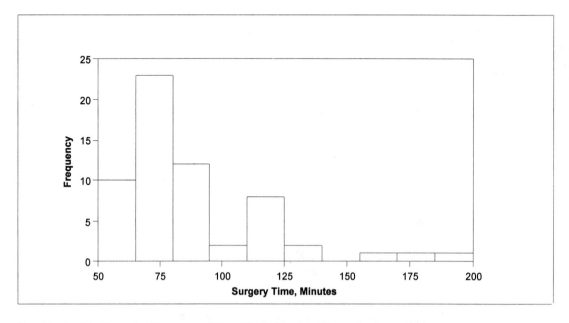

Case Study 6.1 Figure 2 Histogram of Lap Chole Surgery Times for Surgeon A

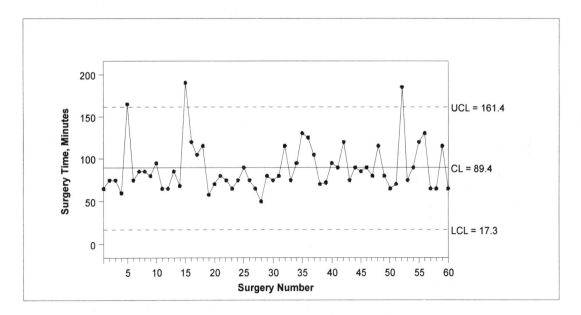

Case Study 6.1 Figure 3 I Chart of Lap Chole Surgery Times for Surgeon A

Since the data are not normally distributed, process capability must be estimated from a probability plot. It would be advantageous if more data were available, but it took one year to get this much data. You could always go back in time to get more data. However, the further back in time you go, the more likely there is to have been a change in the procedure. This would confound the data. Hence, the best must be made of the data available.

The probability plot for surgeon A is repeated in Figure 4 with the best-fit curve. Using the .00135 and .99865 probabilities, the best process capability estimate is [45, 275] minutes. Other information can be gathered from this probability plot. For instance, it may be desirable to schedule for surgeon A to accommodate 95% of A's procedures. To estimate the time needed, a horizontal line is drawn from .95 to intersect the best-fit curve. At the point of intersection, a vertical line is dropped to intersect the x-axis. It does so in Figure 4 at approximately 142 minutes.

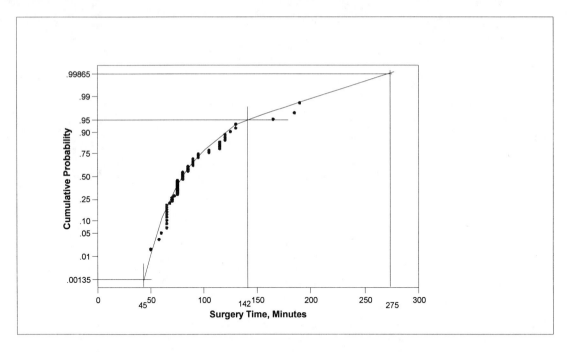

Case Study 6.1 Figure 4 Annotated Probability Plot of Lap Chole Surgery Times for Surgeon A

The Xbar and s chart in Figure 5 profiles the surgeons on their lap chole surgery times. Xbar, *s*, number of surgeries (*n*), control limits, and centerlines for each surgeon are displayed in Table 2.

Note how the control limits step up and down, being wider for the smaller subgroups since less certainty exists there. Furthermore, there are steps in sBar. Both the Xbar and the s charts show evidence of special-cause variation. They are also "in phase," meaning they go up and down together. This is a clear indication that the data are so severely skewed to the right that the averages will not be normally distributed. Also, since the s chart exhibits special-cause variation, the control limits on the Xbar chart are suspect. The distributions of the various surgeons can be thought of schematically, as shown in Figure 6.

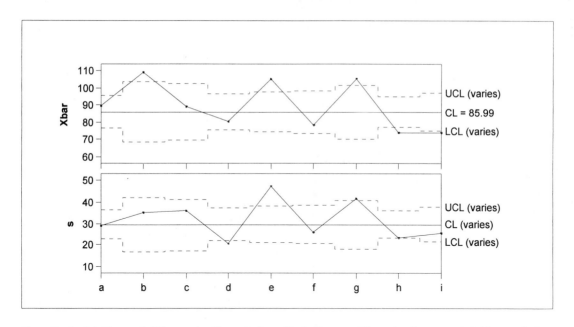

Case Study 6.1 Figure 5 Xbar and s Chart on Lap Chole Surgery Times by Surgeon, 2.5-Sigma Limits

Case Study 6.1 Table 2 Statistics by Surgeon, 2.5-Sigma Limits

	Surgeon								
	A	B	C	D	E	F	G	H	I
Xbar	89.38	109.17	89.10	80.23	105.21	78.34	105.55	73.83	74.00
s	28.78	34.95	35.70	20.40	47.27	25.70	41.33	23.38	25.44
n	60	18	20	48	39	35	22	66	43
UCL(Xbar)	95.54	103.43	102.54	96.67	97.84	98.50	101.77	95.10	97.27
LCL(Xbar)	76.44	68.55	69.45	75.31	74.14	73.48	70.22	76.88	74.71
UCL(s)	36.27	41.76	41.13	37.05	37.86	38.32	40.59	35.96	37.47
sBar	29.47	29.16	29.21	29.44	29.40	29.38	29.25	29.48	29.42
LCL(s)	22.67	16.57	17.29	21.83	20.94	20.44	17.90	23.01	21.37

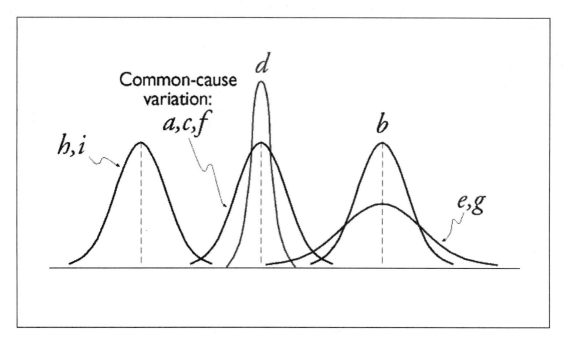

Case Study 6.1 Figure 6 Distributions of the Nine Surgeons

Lessons Learned

The best estimate of process capability comes from a probability plot, particularly if the data are skewed. The probability plot is also helpful to estimate other cumulative probabilities, such as the values below which 95% of the data will fall.

The Xbar and s chart being in phase (i.e., they go up and down together) is an indication that the data are severely skewed to the right.

Management Considerations

The surgery times are skewed to the right by the nature of such procedures. Therefore, some long times can be expected even if the average is low. Since there is a large difference among surgeons, it may be helpful to have the surgeons discuss their approach to the procedure. In this way they may learn from each other and perhaps standardize some aspects of the procedure.

Problems

1.   Referring to Figure 4, compare estimates of the median and 95th percentile of surgery times for surgeons A and B. What is your subjective degree of confidence in the four time estimates and why?

Chapter 7 Using Attribute Data: The c Chart and the u Chart

<u>Using Attribute Data</u>

This chapter and the next address the use of attribute data (also known as "count-type" data) to make attribute control charts. Attribute data are different from measurement data in that it takes only one statistic, the average, to describe a set of data. This is because with attribute data, the standard deviation is a function of the average. Consequently, it takes only a single chart for an attribute control chart analysis—attribute harts do not come in pairs as do the Xbar and s charts.

The following points apply to all attribute analyses:

1. It was noted in Chapter 4 that percentages derived from measurements were variables data. However, quotients involving counted integers, such as a monthly rate of patient falls, are attribute data.
2. Variables data will sometimes be converted to attribute data. An example is to count those cases where systolic blood pressure is greater than 140 mm Hg.
3. The attribute chart centerline is defined as the *process capability*.
4. Attribute charts are used with either time-ordered data (for prediction) or with rational subgroups (for comparison).

The c chart and the u chart are considered in this chapter. The underlying distribution assumed is the *Poisson distribution* (named after Simeon Poisson, who developed the distribution). A typical healthcare application is the study of the monthly count of patient falls.

A *count* does not exist without an *area of opportunity* (or "exposure") in which this count may occur. By custom the area of opportunity is called the *subgroup size*. For a monthly count of patient falls, it is reasonable to express the area of opportunity in which the counted falls occur (i.e., the subgroup size) in terms of the monthly number of patient days.

If the number of patient days each month (i.e., the area of opportunity or subgroup size) is essentially constant, the c chart is used. If the subgroup sizes vary significantly (e.g., the number of patients differs significantly from month to month), a u chart is required rather than the simpler c chart.

The Poisson distribution is skewed to the right (particularly for small average counts), which is reasonable since counts cannot be negative. Just as with variables data, when the distribution becomes too skewed, the usual control chart becomes invalid.

The c Chart

The simplest attribute chart is the c chart, where "c" stands for count. Each plotted point is a c value, which is the count of occurrences for that subgroup. In the case of charting monthly patient falls, the c values are the number of falls for each month, that is, for each subgroup.

The c chart requires that the area of opportunity be essentially constant for each subgroup. By convention, for the c chart this area of opportunity (i.e., this subgroup size) is *always expressed as one unit* where the content of the unit is chosen to suit the application. In the case of charting monthly patient falls, say that there were approximately 1,150 patient days each month. The subgroup size would then be set as one unit of approximately 1,150 patient days.

For the c chart, there are no theoretical restrictions on how large the count may be or on how it is distributed within the area of opportunity. In theory, there could have been 1,500 falls in a month that had 1,150 patient days, and one patient could have had all 1,500 falls and had them all on the same day.

The *Manual on Presentation of Data and Control Chart Analysis* by the American Society for Testing and Materials (ASTM) [1990] suggests that the c chart is most useful when cBar is at least four (otherwise the distribution is too skewed). For cBar less than one, the c chart is not recommended. Some authorities use the sole criterion that cBar be at least five whether or not points are out of control.

If minimum subgroup sizes are not met, subgroups may be combined. For instance, if the average monthly count of patient falls is cBar = 2.5, data can be combined into two-month periods, yielding a satisfactory cBar = 5. The advanced reader may use the Poisson distribution and probability limits [Grant and Leavenworth, 1996] or an adjustment procedure [ASTM, 1990]. (These procedures are beyond the scope of this book.)

If subgroup sizes are larger than needed, there are advantages in splitting them up to make them smaller. If the monthly average of patient falls was 25, it would be well to consider changing to a weekly count and plotting the results on a c chart weekly. In this way you would know much sooner if a special cause of variation occurred.

Unique to the Poisson distribution is the fact that the standard deviation is the square root of the average of the distribution. Let c be the counts of an occurrence and cBar (or $\bar{c}$) be the average of the counts and the centerline (CL) for the chart. Then cBar estimates the average count (per subgroup) for the population, and the estimate of the population 3-sigma of c is

$$3\sigma_c = 3\sqrt{cBar}$$

The upper (UCL) and lower (LCL) control limits are

$$UCL(c) = cBar + 3\sigma_c$$
$$LCL(c) = cBar - 3\sigma_c$$

T-sigma limits may be found by replacing 3 with T from Table 4.1.

Example 7.1 shows the computations of control limits and the c chart for an abbreviated time-ordered data set.

EXAMPLE 7.1

The data in Table 7.1 are counts on medication errors each month where it is assumed that the monthly number of doses (the area of opportunity or subgroup size) is essentially constant. That fixed monthly number of doses is not shown here because it does not enter into the calculations. The c chart is on the monthly count of errors in one unit of a fixed number of doses. The subgroup size for the c chart is always one unit. Say the area of opportunity was approximately 200 doses for each month. Then the c chart would be on the number of errors per unit of 200 doses. The number of errors is not limited by the number of doses since there could be more than one type of error on a dose. For example, a dose may be the wrong medication and it may be delivered at the wrong time and it may be administered by an incorrect method.

3-sigma limits are used for computational simplicity in all examples where the calculation steps are shown.

Table 7.1 Data for c Chart

Month	Number of errors (c)
1	16
2	10
3	15
4	19
5	5

To make a c chart:
cBar = sum of the errors/number of subgroups = 65/5 = 13, which is used for the centerline of the chart. Since cBar is not less than four, the subgroup size (one month) is sufficient.

To calculate the control limits:

$$3\sigma_c = 3\sqrt{cBar} = 3\sqrt{13} = 3(3.61) = 10.8$$

Then
$$UCL(c) = cBar + 3\sigma_c = 13 + 10.8 = 23.8$$
$$LCL(c) = cBar - 3\sigma_c = 13 - 10.8 = 2.2$$

If the calculated value of the LCL had turned out to be negative, it would have been specified as zero since the number of errors cannot be negative. The c control chart is shown in Figure 7.1. □

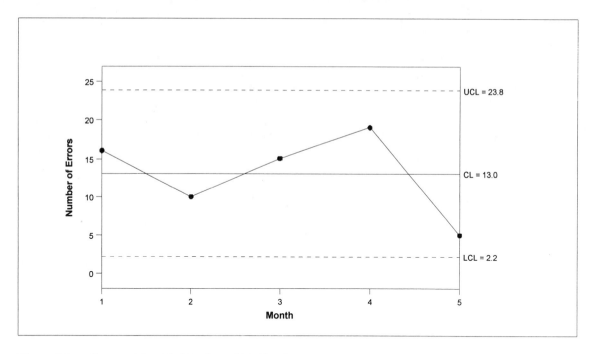

Figure 7.1 c Chart on Monthly Number of Medication Errors

Example 7.1 showed no evidence of lack of control. Had this been the case when the control limits were recalculated after 25 subgroups you would infer that only common-cause variation was present.

With the c chart, the theory and calculations for rational subgroups work the same way as for time-ordered data. For instance, the data from Example 7.1 could have been from five different wards instead of five different months. The restriction that the subgroup sizes must be essentially constant still applies. There will be as many subgroups as there are populations to be compared, and the minimum subgroup size would still be such that cBar would be not less than four.

For standard given with a c chart, the control limits and the centerline (cBar) are used as standard values. For ongoing process control using time-ordered data, these would be projected into the future as shown in Example 7.2. This allows for data collected in real time to be plotted immediately, increasing the chance of discovering the reason for any indications of special-cause variation.

EXAMPLE 7.2

Suppose in Example 7.1 there had been 25 subgroups, so that the calculated control limits could be projected into the future for ongoing process control. Further suppose that the next month 25 errors were found. As shown in Figure 7.2, this is outside the control limits and is evidence of special-cause variation. It would be necessary to immediately investigate to find the cause. The cause may turn out to be that this is the first month that all the errors were reported, or it could be that the number of errors actually increased.

If the latter is the case, it would be advisable to find the reason for the increased number of errors and correct the system so this will not happen again. □

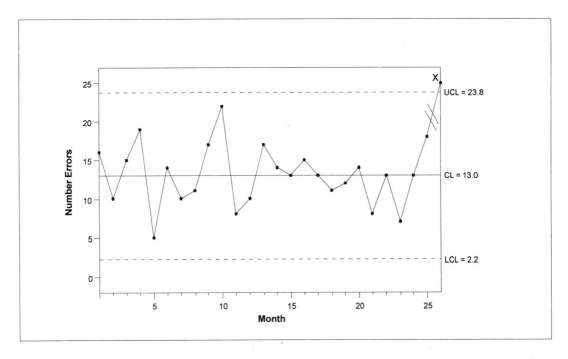

Figure 7.2 c Chart with Standard Given—Monthly Medication Errors for Ongoing Process Control

The c chart may be used out of desperation, where all that is available is a subgroup count and it is reasonable to conjecture that the subgroup size is essentially constant.

The u Chart

For the u chart, the area of opportunity (the subgroup size) need not be constant. In the case of patient falls, the number of patient days might vary grossly from month to month. The subgroups must be compared "on a level playing field." This is done by dividing the subgroup count by the subgroup size, where this size is expressed in whatever "units" are convenient. For patient falls, it might be convenient to use units of 100 patient days, in which case 1,150 patient days in a month would be expressed as 11.5 units of 100 patient days each.

After dividing the subgroup count by the subgroup size in units, the resulting quotient for each subgroup is called a "u" value. These are the values plotted on the u chart, because they may be compared directly. The "u" in u chart stands for "unit" and the chart may be called a "count per unit" chart.

As with the c chart, there are no theoretical restrictions on how high the count may be in a subgroup or on how that count is distributed within the subgroup.

uBar (or $\bar{u}$) is the centerline (CL) of the chart, the estimate for the population count per unit. It is defined as

uBar = (total count from all the subgroups)/(total number of units)

The estimate of the standard deviation of u is a function of uBar. For finding the 3-sigma control limits,

$$3\sigma_u = 3\sqrt{\frac{uBar}{n}}$$

where n is the subgroup size. Then the upper (UCL) and lower (LCL) control limits are

UCL(u) = uBar + $3\sigma_u$

LCL(u) = uBar - $3\sigma_u$

T-sigma limits may be found by replacing 3 with T from Table 4.1.

In accordance with the ASTM [1990], in this book the following guidelines for minimum subgroup size are recommended so that the distribution will not be too skewed. Find uBar; then

1. n should be at least 1/uBar. If not, ignore the subgroup or combine data.
2. A point above the upper control limit is correctly identified as "out of control" only if n is at least 4/uBar. Otherwise, ignore that subgroup or combine subgroups to get the minimum subgroup size needed.

Some authorities recommend n at least 5/uBar as the sole criterion, whether or not points are out of control. (If a point falls above the upper control limit when the subgroup size is greater than 1/uBar but less than 4/uBar, the advanced reader may use an adjustment procedure [ASTM, 1990, p. 60]. For subgroups of size less than 1/uBar, the advanced reader may use Poisson tables and probability limits [Grant and Leavenworth, 1996]. Both of these procedures are beyond the scope of this book.) If the subgroup size is sufficiently large, consider using smaller subgroups with time-ordered data to get an earlier indication when special causes enter the process. Example 7.3 shows the computations of control limits and the u chart for an abbreviated time-ordered data set.

EXAMPLE 7.3

Suppose that in Example 7.1, the number of doses (the area of opportunity or subgroup size) is not relatively constant from month to month. The data from Table 7.1 are repeated here in Table 7.2 with the number of medication doses administered each month given in units of 100 doses.

Table 7.2 Data for u Chart

Month		Number of Errors	Subgroup Size, n, in Units of Hundreds of Doses
1		16	1.50
2		10	1.25
3		15	1.25
4		19	1.25
5		5	1.25
	Sum	65	6.50

To make a u chart:

uBar = sum of the errors/sum of the doses = 65/6.50

       = 10.00 errors per unit of 100 doses

For n = 1.50

$$3\sigma_u = 3\sqrt{\frac{uBar}{n}} = 3\sqrt{\frac{10.00}{1.50}} = 7.75$$

UCL(u) = uBar + $3\sigma_u$
= 10.00 + 7.75 = 17.75

LCL(u) = uBar - $3\sigma_u$
= 10.00 − 7.75 = 2.25

For n = 1.25

$$3\sigma_u = 3\sqrt{\frac{uBar}{n}} = 3\sqrt{\frac{10.00}{1.25}} = 8.49$$

UCL(u) = uBar + $3\sigma_u$
= 10.00 + 8.49 = 18.49

LCL(u) = uBar - $3\sigma_u$
= 10.00 - 8.49 = 1.51

As with the c chart, if a calculated value of the LCL had turned out to be negative, it would have been specified as zero since the number of defects per unit cannot be negative. Note the wider control limits with smaller subgroups because of less certainty when there are less data. This was seen before with the Xbar and s chart.

163

To compute the u values:

Month	u
1	16/1.50 = 10.67
2	10/1.25 = 8.00
3	15/1.25 = 12.00
4	19/1.25 = 15.20
5	5/1.25 = 4.00

These u values are interpreted as 10.67 errors per unit of 100 doses in month 1, 8.00 errors per unit of 100 doses in month 2, and so on. The u values are plotted on the u chart in Figure 7.3. Note the u chart shows no evidence of special-cause variation. ☐

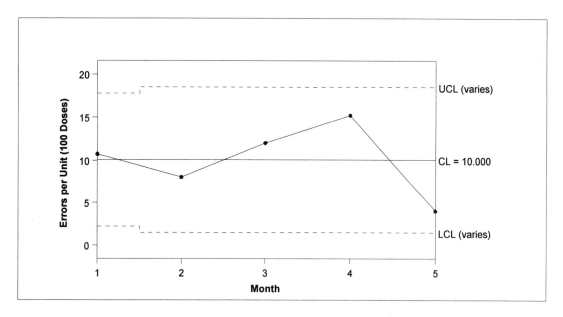

Figure 7.3 u Chart for Monthly Medication Errors per Unit of One Hundred Doses

There was nothing sacred in the choice of 100 doses for a unit in Example 7.3. If you preferred to use units of one dose each, the conclusions from the u chart would be unchanged, as seen in Example 7.4. Choosing to use a different unit simply amounts to "recoding" the data, which has no effect on the outcome of the analysis. Note that this is an entirely different concept from combining or splitting subgroups that was mentioned earlier, which *does* completely change the analysis.

EXAMPLE 7.4

Table 7.3 shows the data from Table 7.2 after recoding the units of the area of exposure from 100 doses per unit to one dose per unit.

164

Table 7.3 Data for u Chart

Month	Number of Errors	Subgroup Size, n, in Units of One Dose
1	16	150
2	10	125
3	15	125
4	19	125
5	5	125
Sum	65	650

To make the u chart:

uBar = sum of the errors/sum of the doses = 65/650 = 0.1000

For $n = 150$

$$3\sigma_u = 3\sqrt{\frac{uBar}{n}} = 3\sqrt{\frac{0.10}{150}} = 0.0775$$

UCL(u) = uBar + $3\sigma_u$
= 0.1000 + 0.0775 = 0.1775

LCL(u) = uBar - $3\sigma_u$
= 0.1000 - 0.0775 = 0.0225

For $n = 125$

$$3\sigma_u = 3\sqrt{\frac{uBar}{n}} = 3\sqrt{\frac{0.10}{125}} = 0.0849$$

UCL(u) = uBar + $3\sigma_u$
= 0.1000 + 0.0849 = 0.1849

LCL(u) = uBar - $3\sigma_u$
= 0.1000 - 0.0849 = 0.0151

To compute the u values:

Month	u
1	16/150 = 0.107
2	10/125 = 0.080
3	15/125 = 0.120
4	19/125 = 0.152
5	5/125 = 0.040

These u values are interpreted as 0.107 errors per unit of one dose in month 1, and so on. The u values are plotted on the u chart in Figure 7.4. Note that the conclusions from the two u charts are unaffected by the recoding of the unit for the area of opportunity. ☐

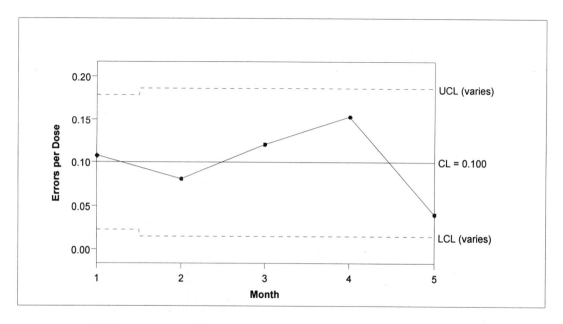

Figure 7.4 u Chart for Monthly Medication Errors per Unit of One Dose

Standard Given

When using standard given with u charts, the only standard value used is for uBar. The control limits are then calculated separately for each subgroup, depending on its area of opportunity.

EXAMPLE 7.5

After 25 subgroups have been plotted, the process in Example 7.4 is found to be a common-cause system, and uBar now may be used for ongoing process control. Month 26 has 20 errors in 100 units of one dose each. Is the process still to be considered a common-cause system?

166

uBar = 0.10 from Example 7.4, so the control limits for month 26 are calculated as follows:

$$3\sigma_u = 3\sqrt{\frac{uBar}{n}} = 3\sqrt{\frac{0.10}{100}} = 0.095$$

UCL(u) = uBar + 3σ_u = 0.10 + 0.095 = 0.195
LCL(u) = uBar - 3σ_u = 0.10 – 0.095 = 0.005

For month 26, u = 20 errors/100 units = 0.20 errors per unit of one dose. As seen in Figure 7.5, this is above the upper control limit, evidence of special-cause variation. An investigation needs to be undertaken to find the cause and hopefully eliminate it. ☐

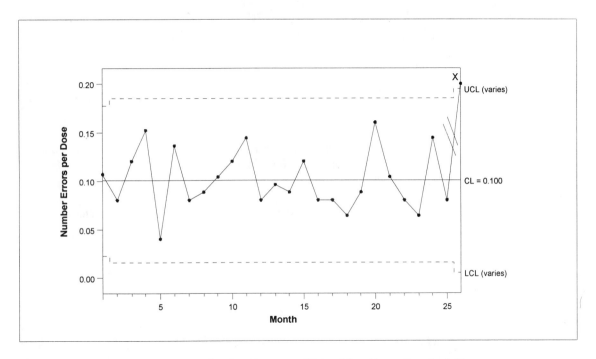

Figure 7.5 u Chart for Monthly Medication Errors per Unit of One Dose, Standard Given

Criteria for Evidence of Nonrandom Influence

For time-ordered data where the purpose is to judge the process for control, only the criterion for one or more points outside the T-sigma limits applies. For time-ordered data where the purpose is improvement, Nelson [1984] suggests only his test 1 (a point outside the control limits), test 3 (six consecutive points, including endpoints, constantly ascending or descending), and test 4 (14 consecutive points alternating up and down) due to the skew of the Poisson distribution. He also suggests that if the distribution is close enough to symmetrical, test 2 (nine consecutive points on the same side of the centerline) may be used. For rational subgroups, only the criterion for one or more points outside the T-sigma limits applies.

167

Chapter Summary

c Chart:

- "c" stands for "count."

- The Poisson distribution is assumed.

- The c chart is a plot of the counts when each subgroup has a constant area of opportunity for the count to occur; for example, the monthly count of patient falls when the monthly number of patient days is essentially constant.

- The subgroup size for the c chart is always one unit where the content of the unit is determined by the application.

- There is no theoretical limit on the count or how it is distributed within the subgroup.

- Formulas for control limits for the c chart:

  $$UCL(c) = cBar + 3\sqrt{cBar}$$
  $$LCL(c) = cBar - 3\sqrt{cBar}$$

- The use of a c chart with cBar less than 1.0 is never recommended, and it is suggested that the minimum subgroup size be large enough that cBar is not less than four. If the subgroup sizes are not large enough, combine subgroups.

- The centerline (cBar) and control limits are used as standard values for standard given.

u Chart:

- "u" stands for "unit."

- The Poisson distribution is assumed.

- The u chart is a plot of count per unit, where the number of units per subgroup may vary.

- The definition of a unit is one of convenience. Coding the data to change the definition of a unit has no effect on the chart interpretation.

- As with the c chart, there is no theoretical limit on the count or how it is distributed within the subgroup.

- Formulas for control limits for the u chart:

  $$UCL(u) = uBar + 3\sqrt{\frac{uBar}{n}}$$

$$LCL(u) = uBar - 3\sqrt{\frac{uBar}{n}}$$

- The subgroup size should never be smaller than 1/uBar, and it is recommended that subgroups not be smaller than 4/uBar. If the subgroup sizes are not large enough, combine subgroups.

- The centerline (uBar) is used as a standard value for standard given.

Problems

1. The number of falls for four consecutive quarters were 92, 85, 78, and 113. No other information is available. Are there indications to suggest nonrandom influence? What value of T should be used for the T-sigma limits? Use T-sigma limits.

2. Continuing with problem 1, more research is done, and it is found that the number of patient days each quarter was 314, 335, 327, and 385, measured in units of hundreds of patient days. With this new information are there now indications to suggest nonrandom influence? (Use T-sigma limits.) Explain.

3. Continuing with problems 1 and 2, do the numbers of patient days for the four quarters show indications to suggest nonrandom influence? Use T-sigma limits. Does this shed any light on the difference in results from the first two problems?

4. For the following medication error data, are there indications to suggest nonrandom influence?

Week	1	2	3	4	5	6	7	8	9	10
Errors	10	8	4	6	3	8	19	7	4	2
Doses	950	950	950	800	800	800	800	950	950	950

5. For the following medication error data, are there indications to suggest nonrandom influence?

Week	1	2	3	4	5	6	7	8	9	10
Errors	1	6	3	5	6	4	9	6	3	2
Doses	700	700	700	500	500	500	500	500	700	500

Computer Supplement (Statit)

EXAMPLE 1

c Chart:
The following data on medication dosage errors per month is saved as sem5.wrk. (Suppose it is possible to have more than one error per dose, for example, incorrect quantity, incorrect time, incorrect drug, and so on.) It is assumed that the number of doses per month is relatively constant.

Month	Number_errors
1	16
2	10
3	15
4	19
5	5

To make a c chart:
    [c] button
        Nonconformities Variable: (click on ▶, Number_errors, Done)
        X labels variable: (click on ▶, Month, Done)
        Control limits
            Upper: Auto (the default)
            Center: (same)
            Lower: (same)
    OK

Note: 3σ limits will be used here for ongoing process control. □

EXAMPLE 2

u Chart:
Modify the data set sem5.wrk on medication dosage errors by adding another variable, Doses, as shown below. (Suppose it is possible to have more than one error per dose, for example, incorrect quantity, incorrect time, incorrect drug, and so on.)

170

Month	Number_errors	Doses
1	16	150
2	10	125
3	15	125
4	19	125
5	5	125

Save the new data set as sem6.wrk.

To change the number of decimal places to 3 and to change the Limit line text:

    Edit > Preferences > QC . . .

        Limit line text: Value and Abb.

        Number of digits after decimal: (type in *3*)

        OK

To make a u chart:

    [u] button

        Nonconformities Variable: (click on ▶, Number_errors, Done)

        Subgroup Sizes variable: (click on ▶, Doses, Done)

        X labels variable: (click on ▶, Month, Done)

        Control limits

            Upper: Auto (the default)

            Center: (same)

            Lower: (same)

    OK

Note: 3σ limits will be used here for ongoing process control. □

Computer Supplement (Minitab)

EXAMPLE 1

c Chart:
The following data on medication dosage errors per month is saved as sem5.xls. (Suppose it is possible to have more than one error per dose, for example, incorrect quantity, incorrect time, incorrect drug, and so on.) It is assumed that the number of doses per month is relatively constant.

Month	Number_errors
1	16
2	10
3	15
4	19
5	5

To make a c chart:
> Stat > Control Charts > C . . .
>> Variable: (click on C2 Number_errors, Select)
>> OK

Note: 3σ limits will be used here for ongoing process control. ☐

EXAMPLE 2

u Chart:
Modify the data set sem5 on medication dosage errors by adding another variable, Doses, as shown below. (Suppose it is possible to have more than one error per dose, for example, incorrect quantity, incorrect time, incorrect drug, and so on.)

Month	Number_errors	Doses
1	16	150
2	10	125
3	15	125
4	19	125
5	5	125

Save the new data set as sem6.

172

To get a list of the plotted points, centerlines, and control limits for up to 16 subgroups to print in the session window, you must use the "brief" command before you make the u chart.
To enable the command language:

Be in the session window (Window > Session). Then
        Editor > Enable Command Language
                at the MTB> prompt in the session window, type in *brief 6*
                hit Enter

To make a u chart:
        Stat > Control Charts > U . . .
                Variable: (click on C2 Number_errors, Select)
                √ at Subgroups in: (click on C3 Doses, Select)
                OK

Note: 3σ limits will be used here for ongoing process control. □

Case Study 7.1   Decreasing Anesthesia Narcotics Discrepancies

Background

Anesthesia narcotics discrepancies are defined in this hospital to include shortages, overages, and incidents of narcotics left unsecured. A quality improvement team has initiated an effort to lower the number of occurrences of anesthesia narcotics discrepancies. A first step is to determine whether historical data on the monthly number of such discrepancies shows special-cause variation. Data have been gathered for the past 24 months.

The Data

(Data are in file ZNARC.) This case study considers only the total number of discrepancies for each of the 24 months given in Table 1. For this historical analysis it was not economically feasible to find the area of opportunity for discrepancies for each month, and it is tacitly assumed to be constant. This sets the stage for a c chart on the monthly count of total discrepancies.

Case Study 7.1 Table 1 Anesthesia Narcotics Discrepancy Data

Observation	Month	Count of Narcotics Discrepancies
1	1	15
2	2	8
3	3	16
4	4	9
5	5	15
6	6	13
7	7	21
8	8	37
9	9	11
10	10	19
11	11	17
12	12	11
13	13	9
14	14	20
15	15	17
16	16	25
17	17	27
18	18	19
19	19	13
20	20	23
21	21	25
22	22	19
23	23	22
24	24	21

Analysis, Results, and Interpretation

Figure 1 is the c chart on monthly anesthesia narcotics discrepancies. No reason could be found, 16 months later, for the point above the upper control limit in the seventh month.

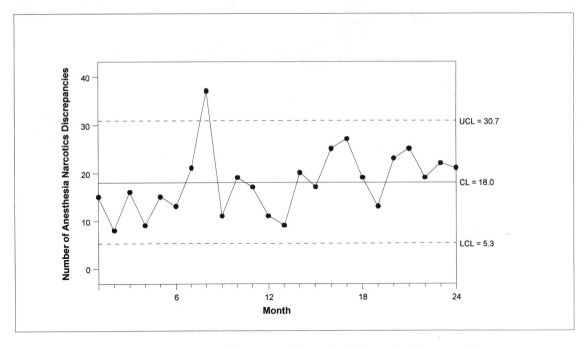

Case Study 7.1 Figure 1 c Chart on Monthly Count of Anesthesia Narcotics Discrepancies

Lessons Learned

It is not unusual to be unable to find special causes of variation with historical data. This is perhaps the best reason to move quickly to plot new data as it becomes available, using control limits from historical data as standard values.

A cursory study of historical data is a wise investment. However, the primary lesson learned from such a study is almost always information on how data should be gathered and analyzed in the future. For example, it would be wise to define a suitable area of opportunity (such as the number of narcotics doses administered) and then measure this for each subgroup.

It would be desirable to have information in a more timely manner than the monthly subgroups provide. The cBar value of 18 here on monthly data implies that a c chart on weekly data would still have a large enough average count. (cBar minimum is not less than 4.)

The real concern is for narcotics shortages, and this information is buried in the present definition of "discrepancies." The initial recording of narcotics shortages or overages should contain more information on the type of narcotic and the magnitude of the discrepancy. If the present method of counting and lumping together all discrepancies is proving to be useful, it would be well to augment that with another set of data exclusively on shortage measurements.

176

Management Considerations

The best way to prevent narcotics discrepancies is to have monitoring methods in place that have accountability and early detection built in.

The data being acquired is not in its most useful form. The effort should be made to get good data promptly so that the likelihood of finding reasons for special-cause variation is increased. Often those with the ability to analyze the data do not have the authority to assign the necessary resources to obtain the necessary data in a timely manner. Help from management may be needed here.

Problems

1. If the data for the 24 months are analyzed with standard given using a standard value of cBar = 22.6, is there still evidence of special-cause variation?

Case Study 7.2   Missing Medications: New Pathway Improvement

Background

A new medication server system has been introduced as a pathway in hopes of improving the availability of medications on the cart when they are to be used. Available historical data provides the number of missing medication events per day. These are occasions where the medications were missing when the RN prepared to give them to the patient. The nurses feel that there has been an improvement since the introduction of the new pathway, but evidence is needed to show that the new pathway is a beneficial special cause of variation.

The Data

(Data are in file ZMISSMED.) Table 1 shows the number of missing medications each day during era 1 (prior to the new pathway) and era 2 (after the introduction of the new pathway). Since the data are count data and the areas of opportunity were not known for each day, c charts are appropriate.

Case Study 7.2 Table 1 Missing Medication Data for Eras 1 and 2 (before and after New Pathway Introduction)

MM = Daily count of missing medications

Day	MM	Era	Day	MM	Era
1	14	1	56	5	1
2	9	1	57	12	1
3	9	1	58	14	1
4	11	1	59	10	1
5	9	1	60	4	1
6	7	1	61	14	1
7	4	1	62	10	1
8	3	1	63	9	1
9	4	1	64	8	1
10	2	1	65	9	1
11	4	1	66	6	1
12	3	1	67	10	1
13	4	1	68	7	1
14	4	1	69	1	2
15	6	1	70	1	2
16	10	1	71	2	2

17	4	1	72	1	2
18	7	1	73	2	2
19	3	1	74	2	2
20	9	1	75	0	2
21	10	1	76	2	2
22	5	1	77	2	2
23	4	1	78	2	2
24	9	1	79	5	2
25	7	1	80	1	2
26	6	1	81	1	2
27	6	1	82	1	2
28	3	1	83	4	2
29	10	1	84	2	2
30	7	1	85	1	2
31	11	1	86	2	2
32	6	1	87	0	2
33	4	1	88	2	2
34	7	1	89	1	2
35	9	1	90	1	2
36	10	1	91	2	2
37	7	1	92	3	2
38	6	1	93	2	2
39	6	1	94	5	2
40	5	1	95	1	2
41	10	1	96	2	2
42	4	1	97	3	2
43	4	1	98	5	2
44	9	1	99	1	2
45	7	1	100	2	2
46	8	1	101	0	2
47	4	1	102	2	2
48	2	1	103	3	2
49	7	1	104	1	2
50	3	1	105	0	2
51	5	1	106	1	2
52	5	1	107	1	2
53	4	1	108	2	2
54	8	1	109	0	2
55	6	1	110	1	2

Analysis, Results, and Interpretation

The improvement of the new pathway is obvious with only a glance at Table 1. All you have to do is find the era in which only single-digit numbers exist for missing medications each day! This is not random chance.

A control chart approach is used for more formal analysis. The best analysis would be to use a p chart (discussed in the next chapter) for the fraction of orders for which medications were missing. However, data on the number of orders is not available, so the best that can be done is a c chart on the count of missing medications per day. This will give a useful portrayal of the data if the number of orders per day is approximately constant.

With a c chart, results from one era may be used as *standard values* for the other era that is then analyzed with *standard given*. The most powerful use of standard given is always to use the time period for which the control limits would be the tightest for the standard values. Accordingly, Figure 1 is the c chart for era 2. Figure 2 shows all of the data (both time periods) using the centerline and control limits from era 2 as standard given. The improvement shown by this portrayal is dramatic.

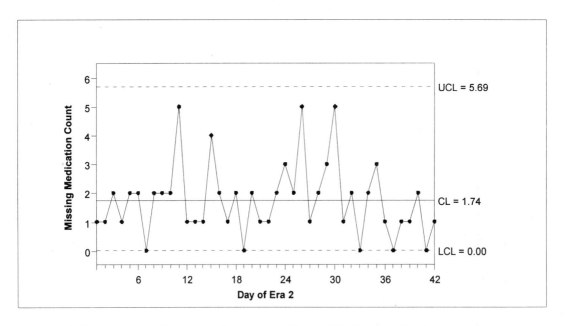

Case Study 7.2 Figure 1 c Chart on Daily Count of Missing Medications, Era 2

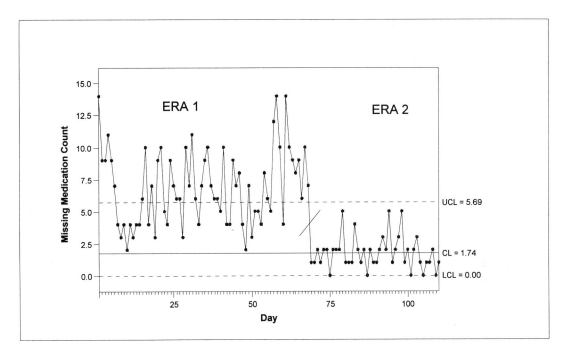

Case Study 7.2 Figure 2 c Chart on Daily Count of Missing Medications for All 110 Days (Both Eras) with Standard Given; Standard Values from Era 2

Lessons Learned

Sometimes the necessary evidence of special-cause variation can be found from patterns in the data, and no further analysis is needed.

The c chart provides a simple tool for use when only raw counts are available. Using a c chart as an approximation for a p chart in this case study was clearly justified since variations in the total number of daily orders could not have altered the conclusion that the new pathway was a profound improvement.

Management Considerations

The new pathway was clearly a big improvement, and it is not difficult to demonstrate that fact.

Problems

1.   If the data for era 2 are analyzed with a standard value of cBar from era 1, is there still evidence of special-cause variation?

2.	Verify that era 1 had 468 missing medications in 68 days and era 2 had 73 missing medications in 42 days. Show that a chart with only two subgroups (the most powerful analysis) confirms that the new pathway was an improvement.

Case Study 7.3   Patient Falls

The concept for this case study was developed with the assistance of Dr. Ray Carey of R. G. Carey and Associates, Park Ridge, IL.

Background

Historical data on patient falls for the last 24 months are to be analyzed as an initial step toward process improvement.

The Data

(Data are in file ZFALLS.) Table 1 shows the number of falls and the number of patient days for each month. The data are counts and the area of opportunity (the denominator, which is the number of patient days per month in this case) is not constant from month to month. Furthermore, a patient may have more than one fall per day, and the numerator (the number of falls) is not part of the denominator. Therefore, the best method of analysis is to make a u chart on falls per patient day.

Case Study 7.3 Table 1 Data on Patient Falls for 24 Months

Observation	Month	Falls	Patient Days (Thousands)
1	1	24	10.426
2	2	36	9.992
3	3	33	10.782
4	4	30	10.767
5	5	27	10.659
6	6	26	10.513
7	7	37	11.093
8	8	26	10.683
9	9	22	10.684
10	10	30	10.794
11	11	28	9.666
12	12	30	9.107
13	13	26	10.085
14	14	25	9.759
15	15	32	10.702
16	16	24	10.384
17	17	29	10.287
18	18	35	9.87
19	19	36	11.293
20	20	30	10.352
21	21	27	11.109
22	22	37	10.842
23	23	27	10.34
24	24	41	10.164

Analysis, Results, and Interpretation

If the area of opportunity were to be constant, a c chart would give exactly the same conclusions as a u chart. Moreover, if the number of patient days per month were approximately constant, the c chart might provide a useful approximation.

Note that the u chart may always be used and that the c chart is never better. The c chart is generally chosen when the area of opportunity is constant because it is simpler. The c chart is also used when the data for the areas of opportunity are not economically available and the assumption that these are essentially constant is reasonable.

To compare the c and u control charts for this case study, the c chart is shown in Figure 1 and the u chart in Figure 2.

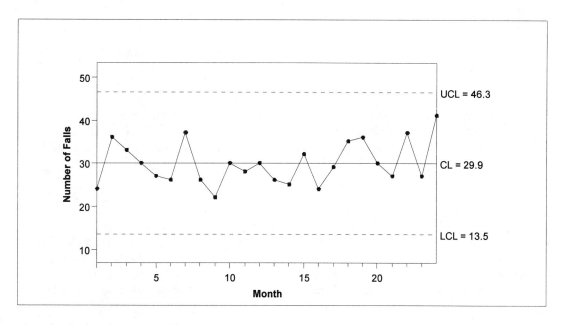

Case Study 7.3 Figure 1 c Chart on Monthly Count of Patient Falls for 24 Months

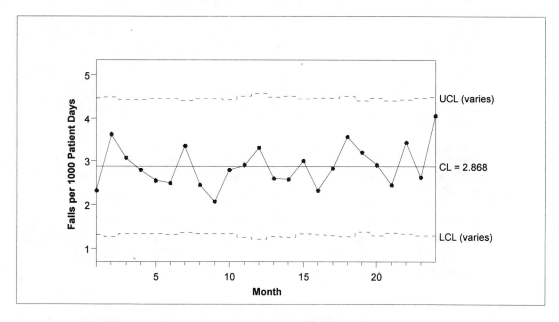

Case Study 7.3 Figure 2 u Chart on Monthly Patient Falls for 24 Months

Lessons Learned

The u chart and the c chart gave the same results in this case, but the u chart should always be used if the area of opportunity can be found. The initial effort at data analysis failed to discover any special-cause variation. Usually, special causes of variation are present and can be found with the proper efforts.

A better look at the historical data is needed, looking at the data by rational subgroups. In parallel with this effort, new data should be obtained on an ongoing basis. The data acquisition should be monitored carefully and the sampling done in such a way as to promote learning, that is, don't be content to look where the light is good—look where the conditions are likely to be bad.

Management Considerations

The control chart shows no evidence of special-cause variation over time. That is not good news for the improvement activity. The hardest type of process improvement is when there are no special causes of variation, because then there is no clue on where to start. It is likely that special-cause variation is present but has not yet been discovered. The most likely reason is that the useful information has been buried by lumping together all interesting avenues for discovery.

Suggestions: (These are management considerations because management will have to spearhead the effort and make the resources available to obtain and analyze the data in the best possible ways.)

1.   Get the knowledgeable people together and ask the circumstances under which they think falls are most likely to occur. Certain shifts? Certain times? Certain days? Certain nurses? Certain physicians? Certain types of patients?

2.   Select the most important variables and the classifications within those variables. Then record for each fall the classification for each variable (e.g., on which of the three shifts did the fall occur).

3.   For each variable, subgroup the data by classification and compare the classifications using the control chart method. This will virtually always flush out opportunities to improve.

4.   Relax the tests used to deliberately promote the possibility of false alarms. Those are a small price to pay for gaining hidden information. The purpose here is not to evaluate the process with respect to special-cause variation but rather to discover clues on how to improve the process.

5.   Have your analyses made on the shortest practical interval. Go to weekly, instead of monthly, intervals. In this case study would there be enough data to allow the use of weekly intervals?

Problems

1. Verify from the data file that there were 349 falls in 125.2 units of 1,000 patient days during the first 12 months and 369 falls in 125.2 units during the second 12 months. Is it better to use a c chart or a u chart to check for special-cause variation between the two 12-month periods? What value of T should be used for the T-sigma limits? What do you conclude?

Chapter 8 Using Attribute Data: The p Chart

Attribute Data

In this chapter, a second type of attribute data is discussed. With this type of data, there is a group of like events, and it is useful to classify some of the events as having a "special" outcome. The "p" value ("p" for proportion) is the "special" outcome proportion found by dividing the count of "special" outcomes (the numerator) by the total count (the denominator). Some examples are listed in Table 8.1.

Table 8.1 Examples of Data Suitable for p Charts

Like Events	"Special" Outcome	p, Proportion
CABG surgery for females over age 65	A death	Mortality proportion
Doses of medication	An incorrect dose	Incorrect medication proportion
Live births after C-section	Vaginal birth after C-section (VBAC)	VBAC proportion
Live births after C-section	Repeat C-section	Repeat C-section proportion
Thoracic deep-wound surgery procedures	Surgical site infection (SSI)	SSI proportion

For each event, the dichotomous outcome is either "special" or "not special." Because there are two names for the outcome, the distribution is called *binomial*. It is common for the special outcome to be considered "unfavorable" but from the examples above it may be seen that this is not universal. It is also common for the "special" outcome to be defined so that p is less than 50%, but this is not always true. To illustrate with the monthly "incorrect medication proportion," the p value for a subgroup of one month is the incorrect medication proportion for the month that may be expressed as a percentage. It is found by dividing the count of incorrect doses (the numerator) by the total dose count for the month (the denominator). The numerator is a subset of the denominator so the number of incorrect doses cannot exceed the total number of doses (i.e., the monthly incorrect medication proportion cannot exceed 100%).

The p values are plotted on a p chart, whether studying time-ordered data for prediction or using rational subgroups for comparisons.

189

The p Chart

The centerline for the p chart is pBar (or $\bar{p}$) defined as

   pBar = (total number of occurrences)/(number of units studied)

This is the estimate of the population proportion. The estimate of the standard deviation of the population proportion is a function of pBar:

$$3\sigma_p = 3\sqrt{\frac{pBar(1 - pBar)}{n}}$$

where n is the subgroup size. The upper (UCL) and the lower (LCL) control limits are

   UCL(p) = pBar + $3\sigma_p$
   LCL(p) = pBar - $3\sigma_p$

T-sigma limits may be found by replacing 3 with T from Table 4.1. In the usual case, pBar is less than 0.5 and the distribution is skewed to the right.

In accordance with ASTM [1990], the following guidelines for minimum subgroup size are recommended so that distribution will not be too skewed. Assuming that pBar is less than or equal to 0.5, find pBar, then

1. n should always be at least 1/pBar. If not, ignore that subgroup or combine data in such a way as to get large enough subgroups.
2. A point above the upper control limit is correctly identified as "out of control" if n is at least 4/pBar. Otherwise, ignore that subgroup or combine subgroups to get the minimum subgroup size needed.

Some authorities recommend n at least 5/pBar as the sole criterion whether or not points are out of control. The advanced reader can use the binomial probability distribution or the adjustment suggested by the ASTM [1990, p. 58].

If time-ordered subgroup sizes are sufficiently large, consider using smaller subgroups to get an earlier indication when special causes enter the process. In the exceptional case where pBar is greater than 0.5, the distribution is skewed to the left, and it is the lower control limit that may be in trouble. Replace pBar with (1 - pBar) in the computation of the minimum subgroup sizes, and replace "above the upper control limit" with "below the lower control limit."

Example 8.1 shows the computations of control limits for the p chart for an abbreviated time-ordered data set.

EXAMPLE 8.1

Table 8.2 has data for five months on monthly counts of medication doses that are incorrect for one or more reasons and the monthly total number of doses. Note for that each dose, the dose is either correct or incorrect. This small example is only to show the computational method.

Table 8.2 Data for the p Chart in Example 8.1

Month	Incorrect Doses	Total Doses
1	16	150
2	10	125
3	15	125
4	19	125
5	5	125

To make a p chart:

pBar = sum of the incorrect doses/sum of the doses = 65/650 = 0.10 = 10% incorrect doses

For $n = 150$

$$3\sigma_p = 3\sqrt{\frac{pBar(1 - pBar)}{n}} = 3\sqrt{\frac{0.10(1-0.10)}{150}} = 0.073$$

UCL(p) = pBar + $3\sigma_p$ = 0.10 + 0.073 = 0.173, 17.3%
LCL(p) = pBar - $3\sigma_p$ = 0.10 - 0.073 = 0.027, 2.7%

For $n = 125$

$$3\sigma_p = 3\sqrt{\frac{pBar(1 - pBar)}{n}} = 3\sqrt{\frac{0.10(1-0.010)}{125}} = 0.080$$

UCL(p) = pBar + $3\sigma_p$ = 0.10 + 0.080 = 0.180, 18.0%
LCL(p) = pBar - $3\sigma_p$ = 0.10 - 0.080 = 0.020, 2%

As with the c and u charts, if a calculated value of the LCL had turned out to be negative, it would have been specified as zero since the percentage of incorrect doses cannot be negative.

As with the Xbar and s chart and the u chart, note the wider control limits with smaller subgroups because of less certainty when there is less data.

To compute the p values:

Month	p
1	16/150 = 0.107 = 10.7% incorrect doses
2	10/125 = 0.08 = 8%
3	15/125 = 0.120 = 12.0%
4	19/125 = 0.152 = 15.2%
5	5/125 = 0.04 = 4%

The p chart is displayed in Figure 8.1.

After the chart has been made, the subgroup sizes should be checked to see if they are large enough. All plotted points were within the control limits, so their required minimum subgroup size of

1/pBar = 1/0.10 = 10 is satisfactorily met. Had there been points outside the control limits, those would have had a required minimum size of 4/pBar = 4/0.10 = 40 (which also would have been met since the actual minimum subgroup size is 125). □

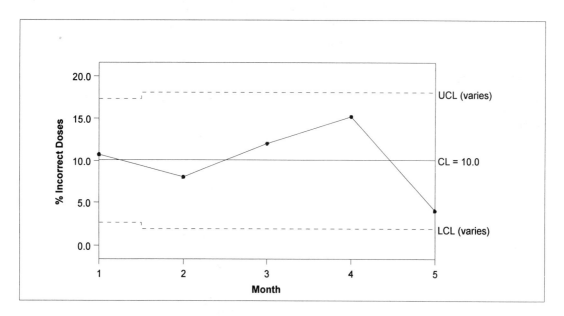

Figure 8.1 p Chart for Percentage of Incorrect Medication Doses

Standard Given

With p charts, the only standard value used is for pBar. The control limits are then calculated separately for each subgroup, depending on its subgroup size.

EXAMPLE 8.2

The process in Example 8.1 is continued for a total of 25 subgroups. It is then found to be a common-cause system with pBar = 0.100 (10.0%). pBar is to be used for ongoing process control. Month 26 has 25 incorrect doses from a monthly total of 100 doses. Is the process still to be considered a common-cause system?

The control limits for month 26 are calculated as follows:

$$3\sigma_p = 3\sqrt{\frac{pBar(1 - pBar)}{n}} = 3\sqrt{\frac{0.10(1 - 0.10)}{100}} = 0.090$$

UCL(p) = pBar + $3\sigma_p$ = 0.100 + 0.090 = 0.190 = 19.0%

192

$$\text{LCL(p)} = \text{pBar} - 3\sigma_p = 0.100 - 0.090 = 0.010 = 1.0\%$$

For month 26, p = 25/100 = 0.250 = 25.0% incorrect doses. As seen in Figure 8.2, this is above the upper control limit and is taken as evidence of special-cause variation. An investigation needs to be undertaken to find the cause and take appropriate action. □

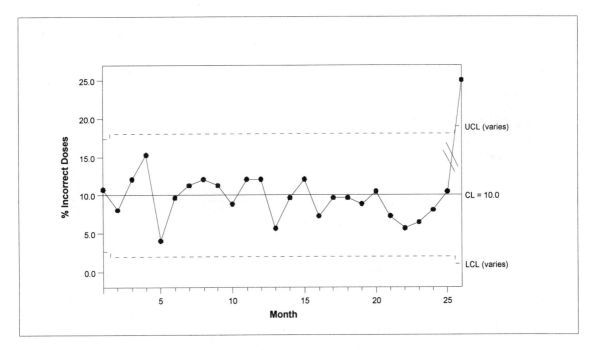

Figure 8.2 p Chart for Ongoing Process Control with Control Limits Established from Months 1 – 25

EXAMPLE 8.3

It is also possible to use limits from one application to use as standard given for another application in order to make a comparison. If ward 1 had a long baseline of data exhibiting only common-cause variation with pBar = 0.050, this value may be used as a standard value to compare ward 2 with ward 1. Ward 2 has collected data for only two months: month 1 with 11 incorrect doses out of 200 doses and month 2 with 16 incorrect doses out of 250. Using pBar = 0.050 from ward 1 as a standard value, the control limits are computed as follows:

For $n = 200$ (month 1):

$$3\sigma_p = 3\sqrt{\frac{pBar(1 - pBar)}{n}} = 3\sqrt{\frac{0.05(1-0.05)}{200}} = 0.046$$

UCL(p) = pBar + $3\sigma_p$
= 0.050 + 0.046 = 0.096 = 9.6%

LCL(p) = pBar - $3\sigma_p$
= 0.050 - 0.046 = 0.004 = 0.4%

For month 1, $n = 200$ and p = 11/200 = 0.055 = 5.5%.

For $n = 250$ (month 2):

$$3\sigma_p = 3\sqrt{\frac{pBar(1 - pBar)}{n}} = 3\sqrt{\frac{0.05(1-0.05)}{250}} = 0.041$$

UCL(p) = pBar + $3\sigma_p$
= 0.050 + 0.041 = 0.091 = 9.1%

LCL(p) = pBar - $3\sigma_p$
= 0.050 - 0.041 = 0.009 = 0.9%

For month 2, $n = 250$ and p = 16/250 = 0.064 = 6.4%.

The control chart is shown in Figure 8.3. Both new points for ward 2 fall within the control limits. Subgroup sizes were then checked and found to be satisfactory. □

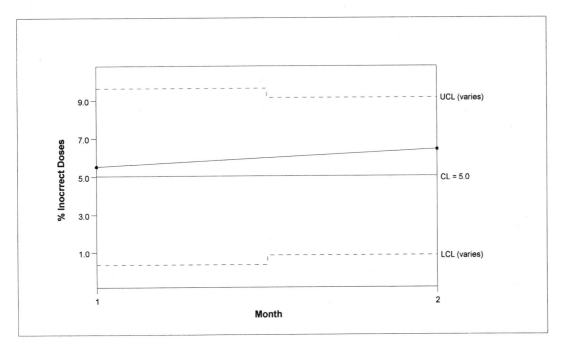

Figure 8.3 p Chart for Ward 2 with Ward 1 as Standard Given

When two or more populations are to be compared, and there is sufficient data from each, the p chart can be used with rational subgroups to compare the applications, as illustrated in Example 8.4.

EXAMPLE 8.4

Data were collected on primary C-section proportions for two physicians, MD 1 and MD 2, for 24 months. Their respective p control charts over that time period are displayed in Figures 8.4 and 8.5. Note that each is a common-cause system. The total number of primary C-sections for the two physicians was 57 and 27, respectively. Also for the two physicians, the total number of births for each that were not following previous C-sections was 310 and 214. Is there evidence of special-cause variation between the primary C-section proportions of the two physicians?

Calculating control limits for a p chart with rational subgroups
$$pBar = (57 + 27)/(310 + 214) = 0.1603$$

For MD 1, $n = 310$

$$3\sigma_p = 3\sqrt{\frac{pBar(1 - pBar)}{n}} = 3\sqrt{\frac{0.1603(1-0.1603)}{310}} = 0.0625$$

UCL(p) = pBar + $3\sigma_p$
= 0.1603 + 0.0625 = 0.2228 = 22.28%

LCL(p) = pBar - $3\sigma_p$
= 0.1603 - 0.0625 = 0.0978 = 9.78%

For MD 1, p = 57/310 = 0.184 = 18.4%

For MD 2, $n = 214$

$$3\sigma_p = 3\sqrt{\frac{pBar(1 - pBar)}{n}} = 3\sqrt{\frac{0.1603(1-0.1603)}{214}} = 0.0752$$

UCL(p) = pBar + $3\sigma_p$
= 0.1603 + 0.0752 = 0.2355 = 23.55%

LCL(p) = pBar - $3\sigma_p$
= 0.1603 - 0.0752 = 0.0851 = 8.51%

For MD 2, p = 27/214 = 0.126 = 12.6%

This information is displayed in the p chart in Figure 8.6, which uses 3-sigma limits for ease of computation. Note that the differences between the proportions of the two physicians can be explained by chance variation. (This is still true even if 1.5-sigma limits are used, as suggested by Table 4.1.) □

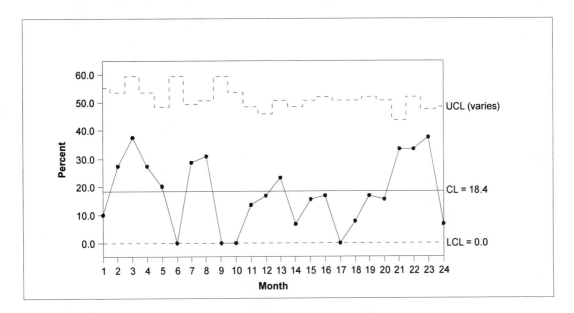

Figure 8.4 p Chart for Primary C-Section Proportion for MD 1

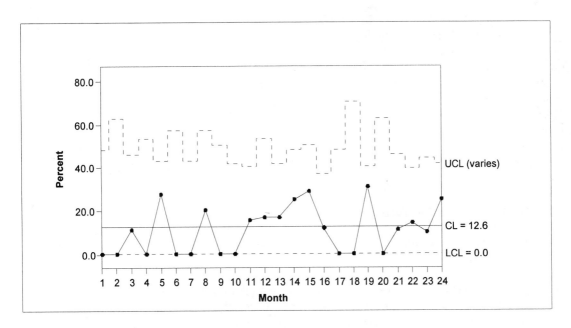

Figure 8.5 p Chart for Primary C-Section Proportion for MD 2

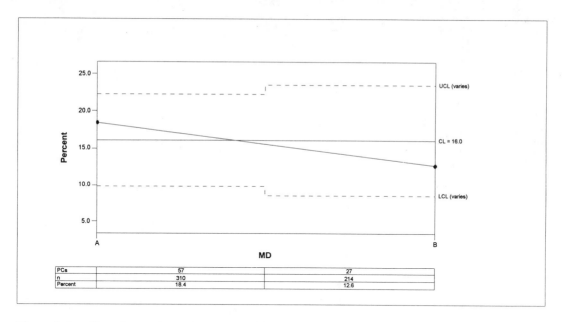

Figure 8.6 p Chart Comparing Primary C-Section Proportions for MD 1 and MD 2

Criteria for Evidence of Nonrandom Influence

For time-ordered data where the purpose is to judge the process for control, only the criterion for one or more points outside the T-sigma limits applies. For time-ordered data where the purpose is improvement, Nelson [1984] suggests only his test 1 (a point outside the control limits), test 3 (six consecutive points, including endpoints, constantly ascending or descending), and test 4 (14 consecutive points alternating up and down) due to the skew of the Poisson distribution. He also suggests that if the distribution is close enough to symmetrical, test 2 (nine consecutive points on the same side of the centerline) may be used. For rational subgroups, only the criterion for one or more points outside the T-sigma limits applies.

p Chart Compared to Chi-Square Analysis

For the reader familiar with classical statistics, it is interesting to note that the p chart is similar in purpose to chi-square analysis.

> When analyzing a set of past samples for lack of control, it is possible as an alternative to a p-chart, to set up the data in the form of a $2 \times n$ contingency table and run a χ^2 test of homogeneity. ... The difference between a p-chart analysis of past data and a χ^2 test lies mainly in the kind of deviation that will lead to rejection of the hypothesis of control or homogeneity. In the former, a single marked deviation will lead to rejection, whereas in the latter, rejection is based primarily upon average departures from hypothetical expectations [Duncan, 1986, pp. 455 – 456].

198

There is a variation of the p chart, the np chart, where the count of "special outcomes" is plotted. The np chart is not covered here because the p chart will *always* suffice.

Chapter Summary

- For the data for a p chart, you can count the nonoccurrences. The number of items studied equals the number of occurrences plus the number of nonoccurrences. (The numerator of p is a subset of the denominator.)

- It is assumed that the data for the p chart follows the binomial distribution.

- Variable (or constant) subgroup sizes may be used.

- 3-sigma control limit formulas:

$$3\sigma_p = 3\sqrt{\frac{pBar(1 - pBar)}{n}}$$

  Then the upper (UCL) and the lower (LCL) control limits are
  UCL(p) = pBar + $3\sigma_p$
  LCL(p) = pBar - $3\sigma_p$

- There are minimum subgroup sizes that should be met.

- The centerline (pBar) is used as a standard value for standard given.

Problems

1.   Make a p chart for the following data on primary C-sections.

Month	PCs	Total w/o Previous C
1	19	143
2	18	121
3	15	124
4	15	126
5	26	118
6	18	125
7	24	120
8	26	130
9	25	133
10	19	139

2.   Make a p chart for the following medication errors.

Month	Doses	Medication Errors
1	17514	5
2	19333	6
3	22657	16
4	21841	3
5	22765	7
6	21026	8
7	26204	7
8	21452	5
9	22878	7
10	21980	8

3.   Make a p chart for the following data on primary C-sections.

Month	PCs	Total w/o Previous C
1	19	143
2	18	121
3	15	124
4	15	126
5	26	118
6	18	125
7	24	120
8	26	130
9	25	133
10	19	139
11	18	136
12	14	111
13	15	126
14	17	132
15	19	121
16	18	127
17	29	124
18	24	128
19	25	134
20	19	137
21	12	146
22	14	121
23	11	167
24	18	143

4.   Make a p chart for the following medication errors.

Month	Doses	Medication Errors
1	21514	7
2	20332	6
3	22659	11
4	23841	11
5	24765	8
6	23026	8
7	24204	9
8	23452	8
9	26878	7
10	23980	8
11	22378	6
12	19348	19
13	22671	6
14	21658	4
15	23557	6
16	22309	7
17	23168	7
18	22382	7
19	24749	8
20	23339	4
21	24748	6
22	24262	9
23	20577	5
24	19491	4

5.   Make a p chart for the following proportions of acute appendectomy perforations. What is the value of T? Use T-sigma limits.

Surgeon	Perforation Count	Procedure Count
1	9	92
2	7	27
3	6	29

6.   For the following infection data, make a p chart. What is the value of T? Use T-sigma limits.

Year	Infection Count	Procedure Count
1	62	412
2	50	401

7.   For the following infection data, make a p chart. What is the value of T? Use T-sigma limits.

Year	Infection Count	Procedure Count
1	42	312
2	36	350
3	24	400
4	16	373

8.   Group the following data by month and by ward and make p charts. What is the value of T? Use T-sigma limits. What do you conclude?

Month	Incorrect Doses	Total Doses	Ward
1	16	150	1
1	10	125	2
2	15	125	1
2	5	125	2

9.   Quoting from "Extra Oxygen Cuts Post-op Infections" in the 20 January, 2000, *USA Today*:

> Simply giving patients a few cents' worth of extra oxygen during and after surgery cuts the risk of infection at the wound site in half, U.S. and European researchers say . . . in today's *New England Journal of Medicine*. . . . 500 patients having colorectal surgery were given either 30% (the standard level) or 80% inhaled oxygen during the operation and for two hours after it. . . . Of the 250 on 80% oxygen, 13 suffered infections, compared to 28 on 30% oxygen.

Was the decrease in percentage infections special-cause variation, or should it be attributed to random chance? Do the recommended 1.5-sigma limits agree with the published report? Would 2-sigma limits give agreement with the published report?

Computer Supplement (Statit)

EXAMPLE 1

p Chart:
Modify the data set sem6.wrk on medication dosage errors by changing the name of the variable
Number_errors to Incorrect_doses to look as follows. (Note that for this example, a dose is either correct or
incorrect.)

Month	Incorrect_ doses	Doses
1	16	150
2	10	125
3	15	125
4	19	125
5	5	125

Renaming a Variable:
To change a variable name
> Right-click on the variable name Number_errors (at top of data column)
> Choose Rename
> Name: type in *Incorrect_doses*
> OK

Save the new data set as sem7.wrk.

To change the number of decimal places to 1
> Edit > Preferences > QC . . .
>> Number of digits after decimal: (type in *1*)
>> OK

In order to make a p chart, choose
> [p] button
>> Rejects variable: (click on ▶, Incorrect_doses, Done)
>> Subgroup Sizes variable: (click on ▶, Doses, Done)
>> X labels variable: (click on ▶, Month, Done)
>> Control limits
>>> Upper: Auto (the default)
>>> Center: (same)
>>> Lower: (same)
>> √ at Display % rejected
> OK

Note: 3σ limits will be used here for ongoing process control. ☐

Grouping Data into Rational Subgroups

It is often desirable to regroup data by rational subgroups.

EXAMPLE 2

For the data in Example 1, suppose that the second and last cases were from ward 2 and all the others were from ward 1. Add a new variable to sem7.wrk called Ward and enter the Ward data as shown below.

Month	Incorrect_ doses	Doses	Ward
1	16	150	1
2	10	125	2
3	15	125	1
4	19	125	1
5	5	125	2

Suppose that you wish to group the data by ward so that you can compare the wards with a p chart. To do that you must add the number of incorrect doses for ward 1 (16 + 15 + 19) and the doses (150 + 125 + 125). The same must be done for ward 2. To do this in Statit Express QC, you need to group the variables.

Grouped Statistics

To group the number of nonconforming units (or nonconformities) for the numerators (Incorrect_doses) and the denominators (Doses):

    Statistics > Grouped Statistics . . .

        Analysis Variables: (click on ▶, Incorrect_doses AND Doses (the variables you want
            grouped), Done)

        Class Variable: (click on ▶, Ward (group by . . .), Done)

        Statistics . . .

            √ at Sum

            remove √ at all other statistics

        OK

These sums will appear in the report window:

STATIT EXPRESS QC REPORT

Ward	Variable	Sum
1	Incorrect_doses	50
	Doses	400
2	Incorrect_doses	15
	Doses	250

Type the sums into a new or existing data file. This is done for you here in the data set sem8.wrk. Make a p chart on this data.

> Edit > Preferences > QC . . .
>> Number of sigmas: 1.5
>> OK
> [p] button
>> Rejects variable:(click on ▶, Total_incorrect_doses, Done)
>> Subgroup Sizes variable: (click on ▶, Total_doses, Done)
>> X labels variable: (click on ▶, Wards, Done)
>> Control limits
>>> Upper: Auto (the default)
>>> Center: (same)
>>> Lower: (same)
>> √ at Display % rejected
>> Sub title: type in *1.5 sigma limits*
> OK

Remember that you will have to change the number of sigmas for the next chart because it will now remain at 1.5 until it is changed again. □

No Standard Given; Standard Given Revisited with p Charts

EXAMPLE 3

Projecting Old Limits into the Future:
Reload data set sem7.wrk. Add a new case:

Month	Incorrect_doses	Doses
6	8	50

Recall that pBar for sem7.wrk is 0.10 = 10%.

Since these limits are to be used into the future for ongoing process control, 3σ limits are appropriate.

In order to make a p chart with pBar = 10% as standard given, choose
[p] button
> Rejects variable: (click on ►, Incorrect_doses, Done)
> Subgroup Sizes variable: (click on ►, Doses, Done)
> X labels variable: (click on ►, Month, Done)
> Control limits
>> Upper: Auto (the default)
>> Center: (click on ▼, Fixed then type in *10*)
>> Lower: Auto (the default)
> √ at Display % rejected
> Sub title: *Limits from Months 1-5*
> OK □

EXAMPLE 4

Using Limits from Another Application:
Clear the old data (File > Clear data; Save changes? No).
Work with data set: sem8.wrk. We wish to use pBar from ward 2 as standard given for ward 1.

Note: As in most of these simple examples, there are not enough data here. This short example is meant to be used for illustrative purposes only.

Sorting Data

To sort by ward:
> Right-click on Ward
> Click on Ascending sort

Save as sem9.

To delete ward 1:
> Click and drag on rows 1 – 3 in the left column to select cases 1 – 3, release

Right-click

Click on Delete

Warning: "This operation cannot be undone."

Click on Continue

Note: Make an ascending sort on Month if no longer in order.

To make a p chart on ward 2 only:

Edit > Preferences > QC . . .

Limit line text: (click on ▼, then click on Abbreviation)

√ at Display data into table below chart

Number of digits after decimal: *2*

Number of sigmas for control limits: *3*

OK

[p] button

Rejects variable: (click on ▶, Incorrect_doses, Done)

Subgroup Sizes variable: (click on ▶, Doses, Done)

X labels variable: (click on ▶, Month, Done)

Control limits

Upper: Auto (the default)

Center: Auto (the default)

Lower: Auto (the default)

√ at Display % rejected

Sub title: *Ward 2*

OK

Note that pBar = 6.00%. This is the "standard given" value of pBar to be used as the centerline on the next p chart.

Open data file sem9.wrk (Save changes? No)

To make a p chart for ward 1 with pBar from ward 2 as standard given (using 3σ limits for ongoing process control):

Delete Ward 2:

Click and drag on 4 – 5 in the left column to select cases 4 – 5, release

Right-click

Click on Delete

Warning: "This operation cannot be undone."

Click on Continue

Note: Make an ascending sort on Month if no longer in order.

To make a p chart on ward 1 using ward 2 as standard given:

Edit > Preferences > QC . . .

> Limit line text: (click on ▼, then click on Abbreviation)
> √ at Display data into table below chart
> Number of digits after decimal: *2*
> Number of sigmas for control limits: *3*
> OK

[p] button

> Rejects variable: (click on ▶, Incorrect_doses, Done)
> Subgroup Sizes variable: (click on ▶, Doses, Done)
> X labels variable: (click on ▶, Month, Done)
> Control limits
> > Upper: Auto (the default)
> > Center: (click on ▼, Fixed, then type in *6*)
> > Lower: Auto (the default)
> √ at Display % rejected
> Sub title: *Ward 1 using Ward 2 as standard given*
> OK □

Computer Supplement (Minitab)

EXAMPLE 1

p Chart:
Modify the data set sem6 on medication dosage errors by changing the name of the variable Number_errors to Incorrect_doses to look as follows. (Note that for this example, a dose is either correct or incorrect.) To rename a variable, click on the variable name, type in the new name, and hit Enter.

Month	Incorrect_ doses	Doses
1	16	150
2	10	125
3	15	125
4	19	125
5	5	125

Save the new data set as sem7.

To get a list of the plotted points, centerlines, and control limits for up to 16 subgroups to print in the session window, you must use the "brief" command before you make the p chart.

Enable the command language if you have not already done so.
Be in the session window (Window > Session). Then
>   Editor > Enable Command Language
>   at the MTB> prompt in the session window, type in *brief 6*
>   hit Enter

In order to make a p chart, choose:
>   Stat > Control Charts > P . . .
>   Variable: (click on C2 Incorrect_doses, Select)
>   Subgroups in: (click on C3 Doses, Select)

>   OK

Note: 3σ limits will be used here for ongoing process control. □

Grouping Data into Rational Subgroups

It is often desirable to regroup data by rational subgroups.

EXAMPLE 2

For the data in Example 1, suppose that the second and last cases were from ward 2 and all the others were from ward 1. Add a new variable to sem7 called Ward and enter the Ward data as shown below.

Month	Incorrect_ doses	Doses	Ward
1	16	150	1
2	10	125	2
3	15	125	1
4	19	125	1
5	5	125	2

Suppose that you wish to group the data by ward so that you can compare the wards. To do that, you must add the number of incorrect doses for ward 1 (16 + 15 + 19) and the doses (150 + 125 + 125). The same must be done for ward 2. Then a p can be made. To do this in Minitab:

> Stat > Basic Statistics > Display Descriptive Statistics . . .
>> Variable: (click on C2 Incorrect_doses, Select, C3 Doses, Select)
>> √ at By variable: (click on C4 Ward, Select)
>> OK

The session window will display N for the number of times ward 1 and ward 2 appear and the respective means as shown below. By multiplying N times the mean, you can get the sum for each ward. For instance, the total number of incorrect doses in ward 1 would be (3)(16.67) = 50. These sums then need to be typed in as new variables, Total_incorrect_doses and Total_doses. This has been done in sem8.

Variable	Ward	N	Mean
Incorrect	1	3	16.67
	2	2	7.50
Doses	1	3	133.33
	2	2	125.00

Load sem8. To make the p chart:

> Stat > Control Charts > P . . .
>> Variable:(click on: C5 Total_incorrect_doses, Select)
>> Subgroup sizes in: (click on: C6 Total_doses, Select)
>> Annotation (click on ▼, Title . . .)
>>> type in *1.5 sigma limits*

OK

S Limits . . .

    Sigma limits positions: (type in *1.5*)

    OK

OK ☐

No Standard Given, Standard Given Revisited with p Charts

EXAMPLE 3

Projecting Old Limits into the Future:
Reload data set sem7. Add a new case:

Month	Incorrect_doses	Doses
6	5	50

Recall that pBar for sem7 is 0.10.
Since these limits are to be used into the future for ongoing process control, 3σ limits are appropriate.

In order to make a p chart with pBar = .10 as standard given, choose:

Stat > Control Charts > P . . .

    Variable: (click on C2 Incorrect_doses, Select)

    Subgroup sizes in: (click on C3 Doses, Select)

    Historical p: (type in *.10*)

    Annotation (click on ▼, Title . . .)

        type in: *Limits from Months 1-5*

        OK

    OK ☐

EXAMPLE 4

Using Limits from Another Application:
Reload sem8. We wish to use pBar from ward 2 as standard given for ward 1.

Note: As in most of these simple examples, there are not enough data here. This short example is meant to be used for illustrative purposes only.

Sorting Data

To sort by ward:

> Manip > Sort . . .
>> Sort column(s): (type in: *C1-C4*)
>> Store sorted column(s) in: (type in *C1-C4*) (or *C8-C11* if you want to keep the original
>>> columns; however, you then need to type in new column headings)
>> Sort by column: (type in: *C4*)
>> OK

Save as sem9.

To delete ward 1:

> Click and drag on rows 1 – 3 in the left column to select cases 1 – 3, release
> Right-click
> Click on Delete Cells

To make a p chart on ward 2 only:

Stat > Control Charts > P . . .
> Variable: (click on C2 Incorrect_doses, Select)
> Subgroups in: (click on C3 Doses, Select)
> Annotation (click on ▼, Title . . .)
>> Title: (type in *Ward 2)*
>> OK
> OK

Note that pBar = 0.06. This is the "standard given" value of pBar to be used as the centerline on the next p chart.

Open data file sem9.

To make a p chart for ward 1 with pBar from ward 2 as standard given (using 3σ limits for ongoing process control):

Delete ward 2:

> Click and drag on 4 – 5 in the left column to select cases 4 – 5, release
> Right-click
> Click on Delete Cells

To make a p chart on ward 1 using ward 2 as standard given:

Stat > Control Charts > P . . .

        Variable: (click on: C2 Incorrect_doses, Select)

        Subgroups in: (click on: C3 Doses, Select)

        Historical p: (type in *0.06*)

        Annotation (click on ▼, Title . . .)

                Title: (type in: *Ward 1 using Ward 2 as standard given)*

                OK

    OK □

Case Study 8.1   C-Sections

The concept for this case study was developed with the assistance of Dr. Ray Carey of R. G. Carey and Associates, Park Ridge, IL.

Background

The administration at a midwestern hospital was concerned about the rate of cesarean sections (C-sections), which seemed to be increasing. It also appeared that some physicians had much higher rates of C-sections than did their colleagues. Data from 19 physicians for the preceding 24 months were tabulated into four classifications. There had been

       479 primary C-sections
       355 repeat C-sections
       254 vaginal births after C-sections (VBACs)
      <u>2,702</u> vaginal births with no previous C-sections
       3,790 total births

These data are shown in Table 1.

Case Study 8.1 Table 1 Live Birth Classifications for 24-Month Period

History	This 24-Month Period		
	Vaginal Births This Period	C-Sections This Period	Total
No Previous C-Section	Only Vaginal Births —Never C-Section 2,702	Primary C-Sections 479	Primary C-Section Candidates 3,181
Previous C-Section	Vaginal Births after C-Sections (VBACs) 254	Repeat C-Sections 355	Repeat C-Section Candidates 609
Total	Total Vaginal Births This Period 2,956	Total C-Sections This Period 834	Total Deliveries This Period 3,790

The Data

(Data are in file ZCSEC.) The C-section figures in Table 1 are counts, not measurements, so they are attribute data. For each delivery either a C-section was performed or it wasn't, so the data are binomial and will be analyzed by p charts. The numbers of occurrences and nonoccurrences can both be counted.

It is important to standardize definitions so that data are always gathered and computed the same way. Doing so, the primary C-section rate is defined here as

Primary C-section rate

$$= \frac{(\text{\# of primary C-sections})}{(\text{total \# who might have had primary C-sections})}$$

$$= \frac{(\text{\# of primary C-sections})}{(\text{total \# of deliveries from mothers w/o previous C-sections})}$$

$$= \frac{(\text{\# of primary C-sections})}{(\text{total deliveries - repeat C-sections - VBACs})}$$

$$= \frac{479}{3,181} = 15.1 \text{ percent}$$

Note that the numerator (479) is a subset of the denominator (3,181), so the numerator (i.e., the number of occurrences or primary C-sections) is limited by the sample size.

The repeat C-section rate is defined here as the proportion of women with a previous C-section who have had another C-section in this 24-month period.

Repeat C-section rate

$$= \frac{(\text{\# of repeat C-sections})}{(\text{total \# who might have had repeat C-sections}}$$

$$= \frac{(\text{\# of repeat C-sections})}{(\text{total \# deliveries from mothers with previous C-sections})}$$

$$= \frac{(\text{\# of repeat C-sections})}{(\text{repeat C-sections} + \text{VBACs}))}$$

$$= \frac{355}{609} = 58.3 \text{ percent}$$

The total C-section rate is calculated as follows, but it must be noted that because it pools primary and repeat C-sections, it may be very misleading.

Total C-section rate

$$= \frac{(\text{total \# of C-sections})}{(\text{total \# who might have had C-sections})}$$

$$= \frac{(\text{total \# of C-sections})}{(\text{total \# of deliveries})}$$

$$= \frac{834}{3,790} = 22.0 \text{ percent}$$

Detailed data by physician for the first month are shown in Table 2 to serve as a representative sample of the complete data set that is in data file CSEC.

Case Study 8.1 Table 2 Detailed Birth Data for the First Month

MD Code	Total Deliveries	Total C-Sections	# of Primary C-Sections	# of Repeat C-Sections	# of VBACs
1	10	1	1	0	0
2	14	3	0	3	3
3	11	3	1	2	1
4	8	2	1	1	1
5	9	1	1	0	1
6	16	5	2	3	0
7	3	0	0	0	0
8	17	7	5	2	0
9	6	0	0	0	0
10	3	2	2	0	0
11	10	3	1	2	0
12	10	1	1	0	1
13	3	1	1	0	0
14	14	3	1	2	1
15	7	1	1	0	1
16	7	3	0	3	0
17	11	4	3	1	3
18	13	3	1	2	0
19	21	2	1	1	3
Totals	193	45	23	22	15

Since the data imply both time order and rational subgroups (by physician), the data must be analyzed both ways. It is possible to initially look at the data subgrouped by physician for

1.   primary C-sections
2.   repeat C-sections
3.   total C-sections (but this may be misleading as will be shown)

It is also possible to initially look at the data over time (by month) for each physician separately and for all physicians combined (pooled) by

1.   primary C-sections
2.   repeat C-sections

Since the comparison of physicians and the combined rates of all physicians over time are of most interest, they will be analyzed first here. However, data analysis is an iterative process. As sources of special-cause variation are found, the analysis must adapt by deleting the affected data to be analyzed separately.

Analysis Using Rational Subgroups

In order to "profile" the 19 physicians to see whether there is a significant difference among them, p charts with rational subgroups by physician are used. The p chart for primary C-section rates for the 24-month period subgrouped by physician is shown in Figure 1. Note that all of the subgroups meet the minimum subgroup size requirement of $4/pBar = 4/0.151 = 26.5$, rounded up to 27.

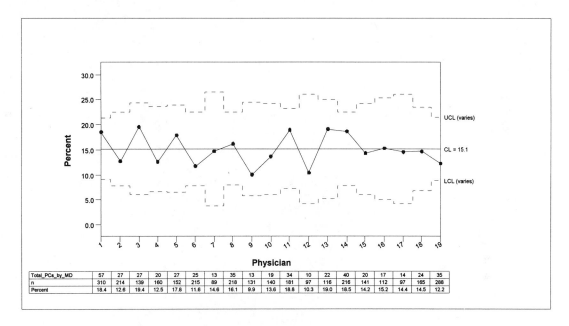

Physician	1	2	3	4	5	6	7	8	9	10	11	12	13	14	15	16	17	18	19
Total_PCs_by_MD	57	27	27	20	27	25	13	35	13	19	34	10	22	40	20	17	14	24	35
n	310	214	139	160	152	215	89	218	131	140	181	97	116	216	141	112	97	165	288
Percent	18.4	12.6	19.4	12.5	17.8	11.6	14.6	16.1	9.9	13.6	18.8	10.3	19.0	18.5	14.2	15.2	14.4	14.5	12.2

Case Study 8.1 Figure 1 p Chart on Primary C-Sections by Physician

Since Figure 1 uses rational subgroups (not time-ordered data), the sole criterion for evidence of special-cause variation is one or more points outside the 3-sigma control limits. There is no evidence of special-cause variation, so it is inferred that the variation in primary C-section rates between physicians for this 24-month period is common-cause variation. The differences among physicians could be explained by chance.

The p chart in Figure 2 again profiles the 19 physicians, this time for their 24-month repeat C-section rate. There is an outage for MD 19. Since the subgroup size of 61 for MD 19 is larger than the minimum of $4/(1 - 0.583) = 9.59$, rounded up to 10, MD 19's 24-month repeat section rate can be interpreted as evidence of special-cause variation; it is so low that it could not reasonably be attributed to random chance. The other 18 physicians are in control with satisfactory subgroup sizes.

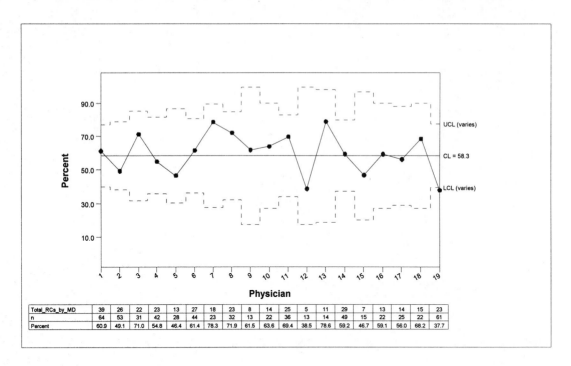

Total_RCs_by_MD	39	26	22	23	13	27	18	23	8	14	25	5	11	29	7	13	14	15	23
n	64	53	31	42	28	44	23	32	13	22	36	13	14	49	15	22	25	22	61
Percent	60.9	49.1	71.0	54.8	46.4	61.4	78.3	71.9	61.5	63.6	69.4	38.5	78.6	59.2	46.7	59.1	56.0	68.2	37.7

Case Study 8.1 Figure 2 p Chart on Repeat C-Sections by Physician

MD 19 will need to be analyzed separately. Remaking the p chart by physician for the 24-month repeat C-section rate without MD 19 results in Figure 3. The other 18 physicians show no evidence of special-cause variation.

220

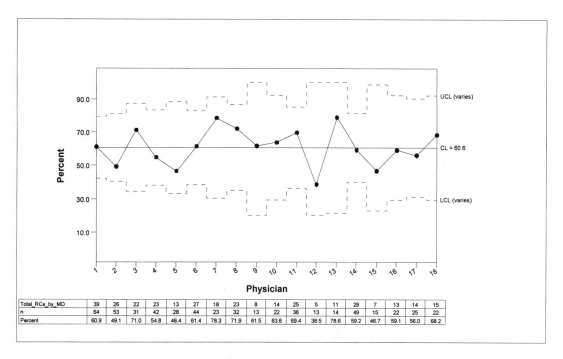

Physician	1	2	3	4	5	6	7	8	9	10	11	12	13	14	15	16	17	18
Total_RCs_by_MD	39	26	22	23	13	27	18	23	8	14	25	5	11	29	7	13	14	15
n	64	53	31	42	28	44	23	32	13	22	36	13	14	49	15	22	25	22
Percent	60.9	49.1	71.0	54.8	46.4	61.4	78.3	71.9	61.5	63.6	69.4	38.5	78.6	59.2	46.7	59.1	56.0	68.2

Case Study 8.1 Figure 3 p Chart on Repeat C-Sections by Physician, MD 19 Excluded

Interest now turns to the profiling of the 19 physicians for their 24-month total C-section rate. The total C-section rate (although widely used) is usually misleading. Indeed, a situation may arise where a hospital might properly state that its total C-section rate has gone down for the time period during when both its primary and repeat rates have increased. For the analysis of an example of this apparent inconsistency, see Appendix 2, "The Perils of Pooling." Pooling data often has the effect similar to having one foot on a hot plate and one foot on a block of ice, so on the average, feeling "fine."

For illustrative purposes, the p chart profiling the 19 physicians for their 24-month total C-section rate is shown in Figure 4. Note that it hides the information that repeat C-sections differ by physician and would hide what the primary and repeat C-section rates are doing. Since total C-section rates may be misleading, no further charts on total C-section rates will be made in this case study.

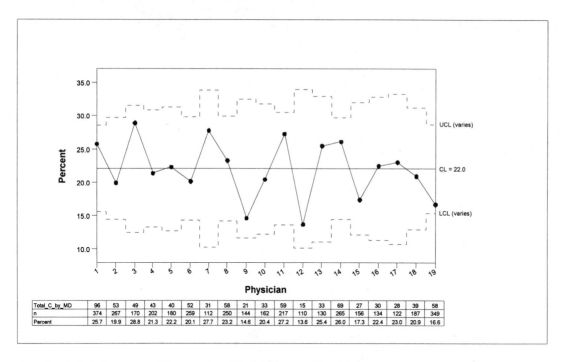

Total_C_by_MD	96	53	49	43	40	52	31	58	21	33	59	15	33	69	27	30	28	39	58
n	374	267	170	202	180	259	112	250	144	162	217	110	130	265	156	134	122	187	349
Percent	25.7	19.9	28.8	21.3	22.2	20.1	27.7	23.2	14.6	20.4	27.2	13.6	25.4	26.0	17.3	22.4	23.0	20.9	16.6

Case Study 8.1 Figure 4 p Chart on Total C-Sections by Physician

Analysis Using Time Order

It is necessary to look at each physician separately over time to see whether that physician is stable (in control) over time, implying only common-cause variation. This is accomplished by the use of a p chart over time for each physician for each of primary and repeat C-section rates. For a given type of C-section, those physicians that are both in control over time and without a special-cause of variation by physician may be pooled. The combined C-section rate of these may then be studied over time.

Analysis of Primary C-Section Rate Using Time-Ordered Data

A separate time-ordered analysis was made of the primary C-section rate for each of the 19 physicians using p charts subgrouped by month. Only MD 13 and MD 18 had points above the upper control limit (Figures 5 and 6). These charts do not meet the required minimum subgroup sizes, but they suggest that these requirements might be met by pooling the data into larger cells (e.g., 2-, 3-, 4-, or 6-month periods).

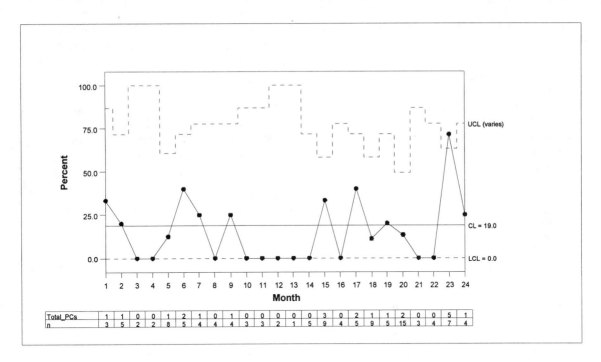

Case Study 8.1 Figure 5 p Chart on Primary C-Section Rates for MD 13 by Month

Month	1	2	3	4	5	6	7	8	9	10	11	12	13	14	15	16	17	18	19	20	21	22	23	24
Total_PCs	1	1	0	0	1	2	1	0	1	0	0	0	0	0	3	0	2	1	1	2	0	0	5	1
n	3	5	2	2	8	5	4	4	4	3	3	2	1	5	9	4	5	9	5	15	3	4	7	4

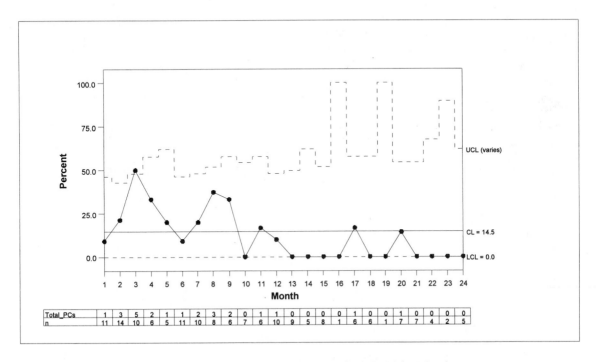

Case Study 8.1 Figure 6 p Chart on Primary C-Section Rates for MD 18 by Month

Month	1	2	3	4	5	6	7	8	9	10	11	12	13	14	15	16	17	18	19	20	21	22	23	24
Total_PCs	1	3	5	2	1	1	2	3	2	0	1	1	0	0	0	0	1	0	0	1	0	0	0	0
n	11	14	10	6	5	11	10	8	6	7	6	10	9	5	8	1	6	6	1	7	7	4	2	5

In order to get the larger subgroup sizes needed for MD 13 and MD 18, the data were grouped into 6-month periods. These larger subgroups give more power in looking at the historical data. However, the price that is paid for the larger subgroups is that there will not be a sufficient number of subgroups to make predictions about future performance.

Figures 7 and 8 show the primary C-section rates for MD 13 and MD 18 using 6-month subgroups with 2-sigma limits (since there are only four subgroups). Figure 7 shows no evidence of special-cause variation for MD 13. Figure 8 suggests a possible special-cause decrease in the primary C-section rate from year 1 to year 2 for MD 18. The p chart subgrouped by year in Figure 9 confirms this suspicion, and the subgroup sizes are quite adequate. Note the 1.5-sigma limits used on this chart due to having only two subgroups.

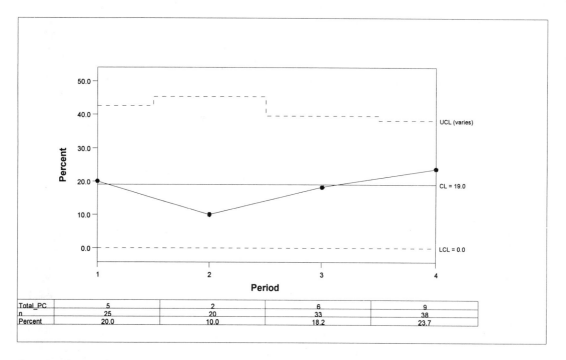

Total_PC	5	2	6	9
n	25	20	33	38
Percent	20.0	10.0	18.2	23.7

Case Study 8.1 Figure 7 p Chart on Primary C-Section Rates for MD 13, by 6-Month Periods, 2-Sigma Limits

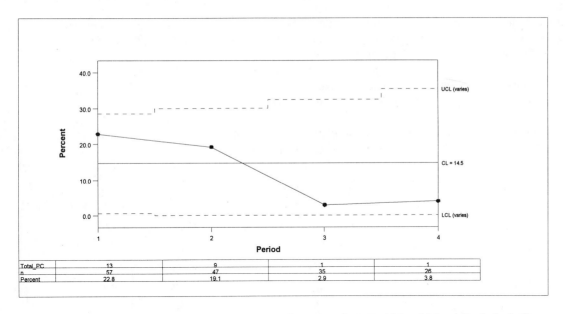

Case Study 8.1 Figure 8 p Chart on Primary C-Section Rates for MD 18 by 6-Month Periods, 2-Sigma Limits

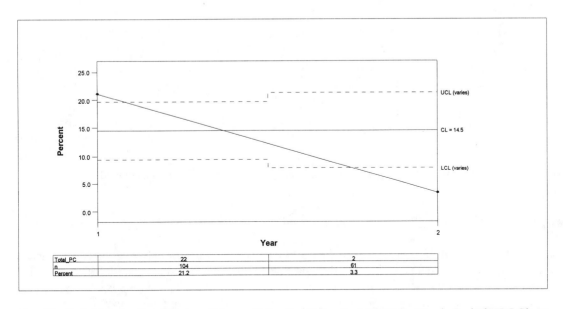

Case Study 8.1 Figure 9 p Chart on Primary C-Section Rates for MD 18 by Yearly Periods, 1.5-Sigma Limits

After examining the primary C-section data in time order for each of the 19 physicians, it is logical to inquire into the analysis of all 19 by pooling their data, month by month. However, the analysis of a set of time-ordered data requires that all of the data arise from what has been called a "single-stream" process. Since all physicians except MD 18 appear to have been stable over time, MD 18 is significantly different from the others. All 19 physicians pooled together do *not* constitute a single stream of data, so it is wrong to analyze them as a single group. MD 18 must be analyzed separately.

The time-ordered primary C-section data for all physicians except MD 18 have been pooled and used for the p chart in Figure 10. No evidence of special-cause variation is seen over time.

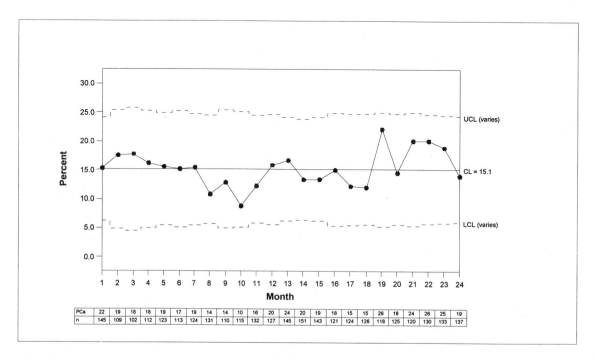

Case Study 8.1 Figure 10 p Chart on Primary C-Section Rates by Month, MD 18 Excluded

Analysis of Repeat C-Section Rate Using Time-Ordered Data

Because of the very small subgroup sizes, valid p charts could not be made for individual physicians, and the repeat C-section rates could only be plotted as run charts. Run charts for MD 7 and for MD 8 showed possible special-cause decreases from year 1 to year 2. (The analysis of the data by year is left as an exercise at the end of the case study.) However, the profiling of physicians for the 24-month repeat C-section rate earlier showed MD 19 to have such a low rate that it was a source of special-cause variation. It would therefore be wrong to pool the 19 physicians together for analysis of their collective time-ordered performance. Therefore, MD 19 must be analyzed separately. The remaining 18 physicians are pooled, and the time-ordered repeat C-section data for the 18 physicians are used for the p chart in Figure 11. The subgroups are of adequate size so a state of control may be inferred.

226

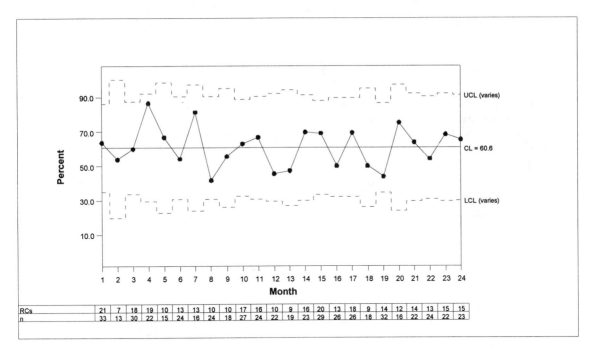

Case Study 8.1 Figure 11 p Chart on Repeat C-Section Rates by Month, MD 19 Excluded

Lessons Learned

Data that are time ordered may also have some underlying rational subgrouping. If so, the data must be analyzed both by rational subgroups and by time order. Then, as sources of special-cause variation are found, the analysis must adapt by deleting the affected data for separate analysis. The rest of the data may be pooled. Note that this requires an iterative process, that is, the data may be analyzed more than once as they are regrouped, separated, or repooled. For instance, MD 19 was a source of special-cause variation when repeat C-section rates were analyzed by physician. Therefore, MD 19 had to be analyzed separately over time. However, since the remaining 18 physicians were part of a common-cause system, they could be pooled over time. Similarly, since MD 18 had a decreasing repeat C-section rate, MD 18's repeat C-section rate had to be analyzed separately. The other 18 physicians, since they were part of a common-cause system and were each separately in control over time, could be pooled over time for repeat C-section rates.

Also, rates that are comprised of separate rates should be dissected and analyzed by the separate rates. Otherwise, the analysis hides information and may be misleading if the individual rates are improperly pooled. Since total C-section rates improperly pool primary and repeat C-sections, the total C-section rate should be avoided. Instead, to better understand the process, primary and repeat C-section rates should be looked at separately.

Managerial Implications

Note that the physicians are not named, only coded. Even in your own application, confidentiality should be enforced. SPC is not a finger-pointing exercise but a statistical tool to gather information from data.

Here, MD 18's primary C-section rate was found to be decreasing, and MD 19 had a lower primary C-section rate. At first glance, these incidents appear to be good. However, the whole picture must be considered. Are MD 18 and MD 19 receiving patients with less complicated pregnancies? Are they endangering more lives? Are they practicing a better protocol? Only the physicians may make those investigations and decisions. The control chart method only looks for special-cause variation over time and physician to physician. Once those special-cause variations are found, it cannot be known whether they are "good" or "bad"—only that they are significantly different. Judicious use of the data and inquiry by knowledgeable people will hopefully bring about discussion among the physicians that will lead to an exchange of ideas and practices. This might then lead to improvements.

Problems

1.   For the data in ZCSEC, make a p chart on primary C-sections for MD 1 by month (Total_PCs/Total_no_previous_C). Does MD 1 exhibit any evidence of special-cause variation for primary C-sections?

2.   For the data in ZCSEC, make a p chart on primary C-sections for MD 2 by month. Does MD 2 exhibit any evidence of special-cause variation for primary C-sections?

3.   Referring to Figure 5, make a p chart for MD 13 on primary C-sections using $k = 8$ quarterly periods. Are the subgroup sizes large enough? Do the results suggest nonrandom influence?

4.   Referring to Figure 6, make a p chart for MD 18 on primary C-sections using $k = 8$ quarterly periods. Are the subgroup sizes large enough? Do the results suggest nonrandom influence?

5.   For the data in ZCSEC, make a run chart for MD 7 and a run chart for MD 8 for repeat C-sections (Total_RCs/Total_previous_C). How can you determine whether there is special-cause variation from year 1 to year 2 for either of them?

6.   For the data in ZCSEC, make a p chart for repeat C-sections for MD 7 and one for MD 8 by year using 1.5-sigma limits. Make p charts similar to Figures 3 and 11 eliminating all physicians who showed special-cause variation over time or by rational subgroups.

Case Study 8.2   Decreasing the Medication Error Rate

Background

Medication doses that are incorrect for any reason are *incorrect doses*, but are more frequently called *medication errors*. The medication error rate expresses these as a fraction of the total number of doses. This fraction is sometimes expressed as medication errors per thousand doses or as medication errors per hundred doses (percent). An ongoing effort has been made over the past two years to decrease the medication error rate. Now a control chart analysis is to be made to determine whether there has been improvement.

The Data

(Data are in file ZMEDERRS.) Table 1 shows the number of reported medication errors and the number of doses for each of the last 24 months, where a dose that was incorrect for any reason is called a medication error. Either a dose is an error or it isn't. The numerator (the number of incorrect doses) is a subset of the denominator (the total number of doses), implying that the medication error rate cannot exceed 100%. This makes the p chart the appropriate control chart.

Case Study 8.2 Table 1 Number of Medication Errors (MedErrs) and Number of Doses for 24 Months

Month	Doses	MedErrs
1	225143	58
2	213325	67
3	246599	130
4	238411	118
5	247650	84
6	230267	89
7	242045	98
8	234524	86
9	268785	72
10	239807	89
11	223788	61
12	203486	53
13	226713	65
14	216582	48
15	235579	62
16	223099	72
17	231685	71
18	223823	78
19	247490	80
20	233398	45
21	247483	63
22	242622	93
23	205776	52
24	194911	40

Analysis, Results, and Interpretation

A p chart (Figure 1) shows the monthly fraction errors. Note that the minimum subgroup size of 4/pBar = 12,500 is met. The p chart is out of control and indicates the strong possibility of a downward trend over the two years. One way to get a sharper look at this would be to compare the first year to the second in a p chart with two subgroups.

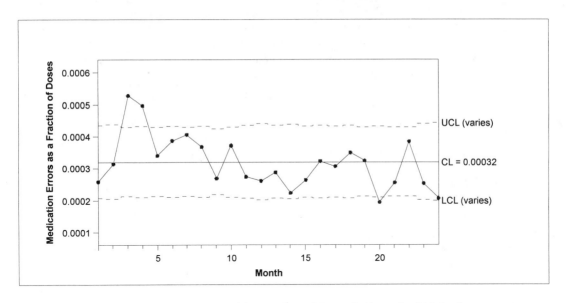

Case Study 8.2 Figure 1 p Chart on Monthly Fraction of Doses in Error for 24 Months

Lessons Learned

The apparent improvement was probably the result of a concerted effort. That effort should be studied carefully and used as a starting point, not ignored. As described in the following section, there is much more that can be done using SPC methods if serious efforts are to be made for process improvement.

Management Considerations

Since the early subgroups showed special-cause variation with high rates of errors and most recent subgroups showed special-cause variation with low rates of errors, it appears that the error rate is decreasing. However, the analyses made above barely scratch the surface of what can be done if a serious effort is to be made to decrease medication errors. To be effective such an effort would require full management participation (not merely management approval). A two-part program would include the following:

1.   A p chart should be continued for ongoing process control, but with weekly subgroups, not monthly. (The reader should verify that the subgroup sizes would, indeed, be adequate.) It would be wise to use both 3-sigma control limits and 2-sigma warning limits. Since process improvement is the objective, if the 3-sigma limits don't quickly flush out any special-cause variation, responding to the 2-sigma limit indications would definitely be good practice.

2.   Seek the help of those people who are most knowledgeable, that is, those closest to the operations. Meet with them and tally their opinions on the factors most likely to affect

medication errors. These are the variables to be broken into rational subgroups for comparisons using p charts. For example, compare the three shifts, compare weekends to weekdays, compare different work areas, compare early morning and late afternoon, etc. When even a hint of special-cause variation is found, look for a reason. Wild goose chases are cheap compared to failure to improve.

Problems

1.   Make a p chart with $k = 2$ for the first year versus the second year.

Case Study 8.3   Acute Appendicitis: Ultrasound Predictive Errors

Background

Specificity and *sensitivity* are the measures of the excellence of a test that are most commonly used. However, specificity and sensitivity do not address the question that is in the patient's mind: "Given that the test says I am diseased, what is the probability that the test is wrong?" Or, "Given that the test says I am normal, what is the probability that the test is wrong?" These two questions are answered, respectively, by the *positive predictive error rate* and the *negative predictive error rate*.

Test results may be correct or they may be incorrect. The radiologist is an integral part of the test. Some radiologists will do better than others. Some radiologists are specialists and some are non-specialists; you would expect the specialists to have the better track record. This case study profiles seven radiologists by their observed positive predictive error rate and again by their observed negative predictive error rate.

The Data

(Data are in file ZULTRA.) Table 1 gives data for profiling the seven radiologists on their error rates. The first three radiologists are radiology specialists; the last four are non-specialists. Table 2 gives the data for profiling the two radiologist groups.

Case Study 8.3 Table 1 Acute Appendicitis: Ultrasound Predictive Errors by Radiologist (RAD)

RAD	FalsPos	TestPos	FalsNeg	TestNeg
1	1	8	1	15
2	4	12	0	35
3	2	8	0	23
4	3	5	2	26
5	5	7	1	17
6	2	6	1	16
7	2	4	1	11

Group	FalsPos	TestPos	FalsNeg	TestNeg
Specialists	7	28	1	73
Non-specialists	12	22	5	70

The observed positive predictive error rate for a data set is the number of false positive (FalsPos) test results divided by the total number of positive (TestPos) test results. Since the numerator is a subset of the denominator, the p chart is used.

Analysis, Results, and Interpretation

Figure 1 is a p chart with 2.5-sigma limits profiling the seven radiologists on their positive predictive error rate. Note the demarcation between the two groups of radiologists. All points fall within the control limits, so the subgroup sizes are adequate.

As expected, the specialists appear to be doing a better job. The critical question is, "Might the apparent superior performance of the specialists simply be random chance, or is there really special-cause variation between the two groups?"

Figure 2 is a p chart with 1.5-sigma limits profiling the two radiologist groups on their positive predictive error rate. This chart confirms the superiority of the specialists with respect to their observed positive predictive error.

Similarly, Figures 3 and 4 profile the seven radiologists and the two radiologist groups with respect to their negative predictive error rates (the number of false negative (FalsNeg) test results divided by the total number of negative (TestNeg) test results). There is nothing to show special-cause variation here—but that does *not* give any evidence that no difference exists.

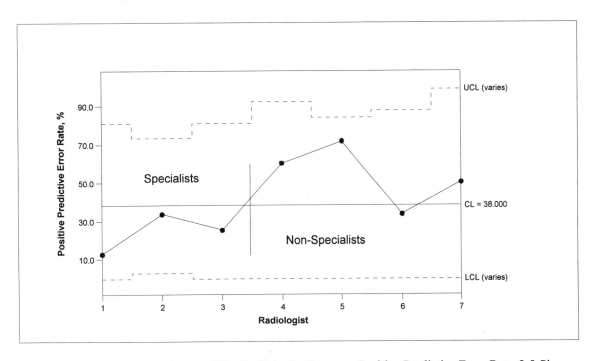

Case Study 8.3 Figure 1 p Chart Profiling Radiologist Group on Positive Predictive Error Rate, 2.5-Sigma Limits

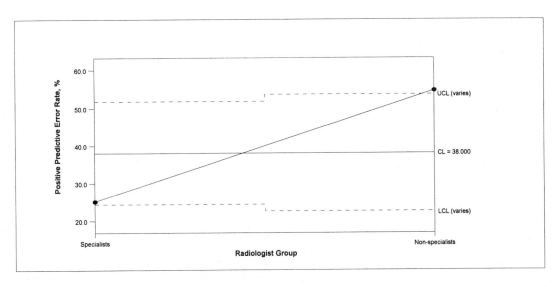

Case Study 8.3 Figure 2 p Chart Profiling Radiologist Group on Positive Predictive Error Rate, 1.5-Sigma Limits

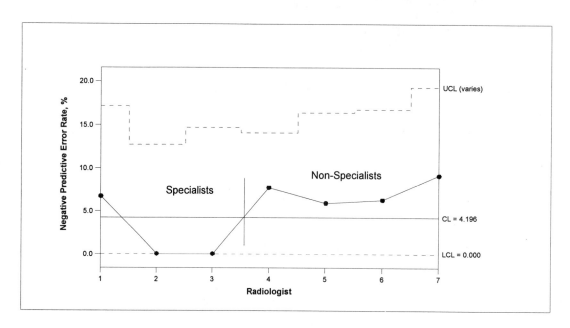

Case Study 8.3 Figure 3 p Chart Profiling Radiologists on Negative Predictive Error Rate, 2.5-Sigma Limits

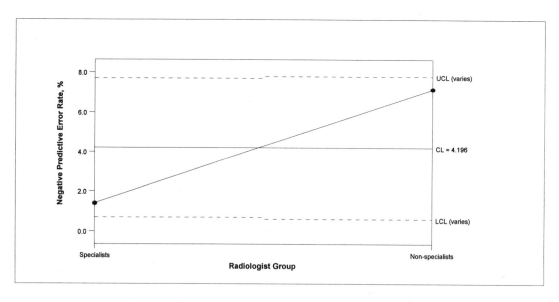

Case Study 8.3 Figure 4 p Chart Profiling Radiologist Group on Negative Predictive Error Rate, 1.5-Sigma Limits

Lessons Learned

With the positive predictive error rates, the p chart on individual radiologists had insufficient power to detect any special-cause variation. By subgrouping the seven radiologists into specialists and non-specialists, the larger subgroup sizes provided the increased power needed to detect the special-cause variation.

With negative predictive error rates, using the two groups did not provide the needed power to detect special-cause variation. It may be present, but there is not sufficient evidence for a valid evaluation.

Note: Specificity and sensitivity are *intrinsic* measures of the excellence of a test, that is, they have the desirable property of being independent of the prevalence of the disease. A limitation on the usefulness of predictive error rates is that they are formally applicable only to the data set for which they are calculated because they depend on the prevalence of the disease that will vary from data set to data set. You may work around the limited applicability of predictive error rates when they are to be used for comparisons by adjusting them to take into account the prevalence of the disease. This refinement had no effect on the conclusions in this case study.

Management Considerations

Should the question be asked whether only specialists should be used for ultrasound tests?

Problems

1.   Verify that the control charts for negative predictive error rate are correct.

Case Study 8.4 Acute Appendicitis: Perforation Rates by Surgeon

Background

Perforation in an acute appendectomy is a complication of considerable concern. Perforation rate is defined as the number of perforations divided by the number of acute appendectomies. One step toward process improvement, in this and other surgical procedures, is to profile surgeons so they may observe the results and learn from any special-cause variation that may be found.

Although the perforation rate may be largely out of the surgeons' control, this case study looks at whether surgeons who perform fewer appendectomies have higher perforation rates.

The Data

(Data are in file ZPERF.) Table 1 provides the data to profile surgeons by perforation rate for the past year. To determine whether surgeons who perform fewer appendectomies have higher perforation rates, the surgeons have been divided into two groups: those who performed fewer than 20 procedures and those who performed 20 or more. Table 2 provides the data from the past year to profile these two surgeon groups by perforation rate. The p chart may be used to profile the surgeons, since the numerator count (the number of perforations) is a subset of the denominator count (the number of acute appendectomies).

Case Study 8.4 Table 1 Acute Appendectomy Perforation Rate Data by Surgeon

Surgeon	Perforation Count	Procedure Count	Perforation Rate
1	2	22	0.091
2	4	30	0.133
3	3	20	0.150
4	1	21	0.048
5	2	4	0.500
6	4	18	0.222
7	3	12	0.250
8	1	7	0.143
9	2	8	0.250

Case Study 8.4 Table 2 Acute Appendectomy Perforation Rate Data by Surgeon Group

Surgeon Group	Group Perforation Count	Group Procedure Count
Surgeons with < 20 procedures	12	49
Surgeons with 20 or more	10	93

Analysis, Results, and Interpretation

Profiling surgeons, a p chart (Figure 1) with 2.5-sigma limits appears to detect only common-cause variation among the nine surgeons, but the subgroups are too small for the results to be conclusive.

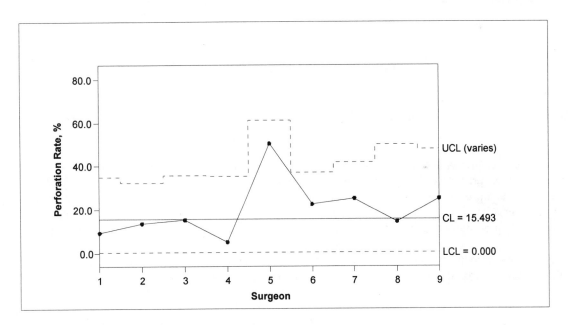

Case Study 8.4 Figure 1 p Chart Profiling Surgeons by Perforation Rate, 2.5-Sigma Limits

More power to detect special-cause variation may be obtained by pooling data to obtain larger subgroups. Figure 2 is a cross-plot or *X-Y* plot to see whether surgeons who perform fewer acute appendectomies appear to have higher perforation rates. This does seem to be the case, but statistical evidence is needed.

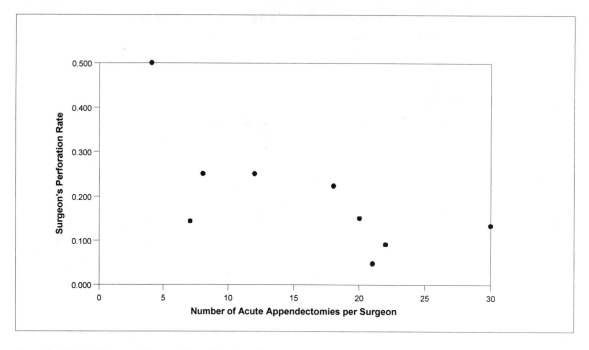

Case Study 8.4 Figure 2 Cross-Plot of Perforation Rate as a Function of Number of Procedures Performed

To determine whether surgeons who perform fewer appendectomies can be shown to have higher perforation rates, the surgeons have been divided into two groups: those who performed fewer than 20 procedures and those who performed 20 or more. The evidence sought is provided by Figure 3, a p chart using 1.5-sigma limits, where it is seen that special-cause variation exists between the surgeon group with fewer than 20 procedures and the group with 20 or more.

Note that the subgroup sizes are large enough for the control chart results to be believed. Yes, surgeons with fewer procedures had significantly higher perforation rates in this study.

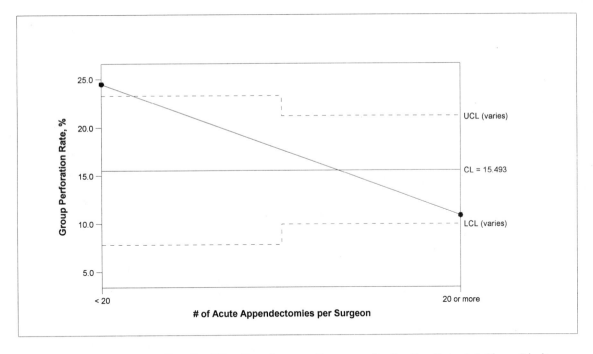

Case Study 8.4 Figure 3 p Chart Profiling Two Surgeon Groups on Perforation Rate, 1.5-Sigma Limits

Lessons Learned

The key to process improvement is discovering special-cause variation. When the first control chart fails to detect this, it may be because the data has not been subgrouped optimally or because there is simply not enough data. A process without special-cause variation is theoretically possible, but almost never found until improvements have been made. Success in discovering special-cause variation takes expert knowledge of the process to know how to subgroup the data and statistical knowledge to know how to maximize the power of the control chart method.

Management Considerations

Since the perforation rate may be largely out of the surgeons' hands, considerable care is needed to discover the reason for the special-cause variation shown in Figure 3.

Problems

1.   Why is pBar the same in Figures 1 and 3? What is the minimum subgroup size to believe an out-of-control point for that subgroup? What is the absolute minimum subgroup size that should be used? Is this latter minimum met in the above figures?

Chapter 9 Transformations (An Advanced Topic)

The Need for Transformations

Variables control charts require that the data be near-normal. Many processes, by their very nature, are not near-normal. For instance, surgery times may not be. There is some minimum time that the surgery will take, and it obviously cannot take less than 0 minutes. However, there is no real upper bound. This often yields data that are very skewed to the right. As an example, consider the case study on surgery times after Chapter 6. Surgeon A's data were found to be highly skewed to the right. (In fact, all of the surgeons' data were.)

It is common practice in statistics to transform the data if necessary so that the data meet the necessary assumptions. This is often accomplished by choosing the correct power (or logarithm) of the data to make it meet the assumption.

Let X be the variable being studied. Initially, the variable is to the first power. It is possible to take higher powers of the data—X^2 (squared), X^3 (cubed), X^4 (fourth power), and so on. It is also possible to take smaller positive powers—$X^{1/2}$ (square root), $X^{1/3}$ (cube root), $X^{1/4}$ (fourth root), and so on. Since $X^0 = 1$, the $\ln(X)$ (natural log) is used instead. (This is appropriate since X^t behaves much like the logarithm function for t close to 0 [Clevelend, 1993].) It is also possible to take negative powers. (Recall that taking X to a negative power is taking 1 over X to the positive power.) These include X^{-1} ($1/X$), X^{-2} ($1/X^2$), and so on.

In order to determine the best power transformation to make the data near-normal, the following guidelines are suggested, recognizing it will be a trial-and-error process:

1. Plot the original data on a probability plot.
2. If the data appear skewed, try a lower power.
3. Make a probability plot on the transformed data:
   - If the probability plot looks better (less skewed), take lower powers of the original data.
   - If the probability plot looks worse (more skewed), take higher powers.
4. Continue until the probability plot looks reasonably straight.

EXAMPLE 9.1

Consider the 60 surgery times of surgeon A. (See Case Study 6.1, Table 1.) The probability plot of surgeon A's data is shown in Figure 9.1 (repeated from Case Study 6.1 Figure 1). The probability plot is severely convex upward, indicating that the distribution of X, the surgery times, is highly skewed to the right. These original data are the X (or X^1) values. A lower power of X is suggested. Try the square root of X, which is X

to the 1/2 power, $X^{1/2}$ or $X^{0.5}$. To illustrate the calculations involved, the first two surgery times from surgeon A are 65 and 75 minutes. Taking the square root of the surgery times yields 8.06 and 8.66 for the first two. The probability plot (Figure 9.2) is still convex upward, only not as much.

Try $\ln(X)$. Taking the natural log of 65 and 75 yields 4.17 and 4.32, respectively. The probability plot (Figure 9.3) shows less upward convexity.

Try $X^{-1/2}$ or $X^{-0.5}$. Taking 1 over the square root of 65 and 75 yields 0.124 and 0.115, respectively. The probability plot (Figure 9.4) is almost a straight line.

Try X^{-1}, the reciprocal of X. Taking 1 over 65 and 75 yields 0.0154 and 0.0133, respectively. The probability plot (Figure 9.5) looks satisfactorily straight.

Try X^{-2}. Taking 1 over the 65 and 75 squared yields 0.0002367 and 0.0001778, respectively. The probability plot (Figure 9.6) looks as though the straight-line fit is poorer.

X^{-1} will transform the original data into the needed near-normal data. The original surgery times, the X values, were expressed in terms of minutes per procedure. X^{-1} is the reciprocal of X, the number of procedures per minute (which could also be expressed as the number of procedures per hour).

This example is carried out in full in the case study on surgery times at the end of this chapter. □

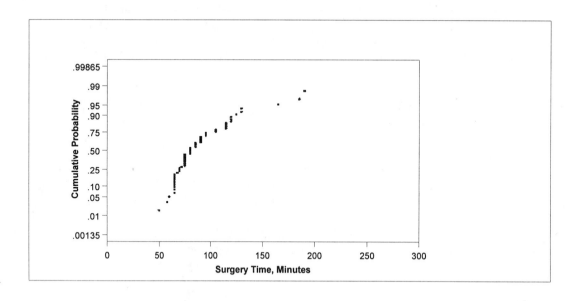

Figure 9.1 Probability Plot of Surgeon A's Surgery Times (Original Data)

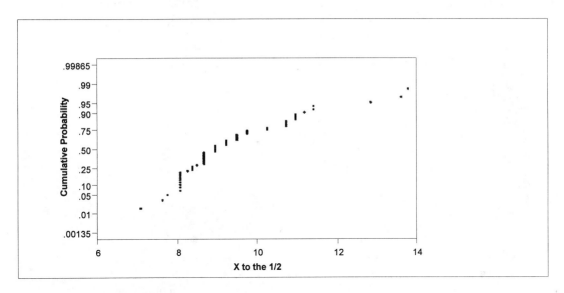

Figure 9.2 Probability Plot of X to the ½ or $X^{0.5}$

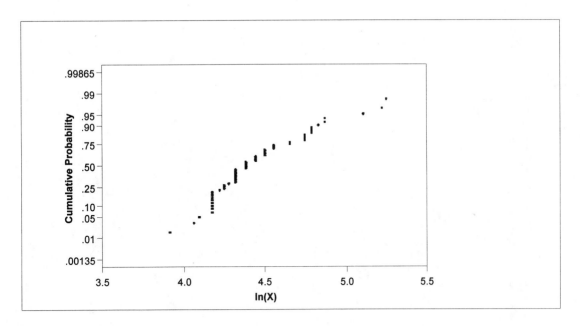

Figure 9.3 Probability Plot of ln(X)

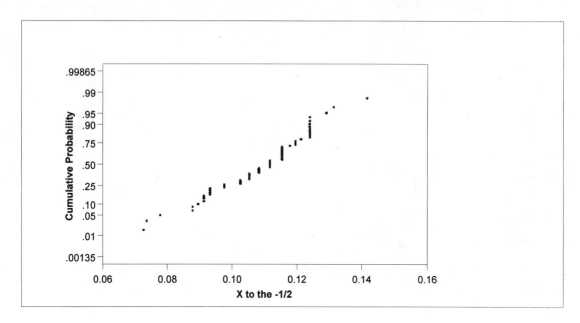

Figure 9.4 Probability Plot of X to the -½ or $X^{-0.5}$

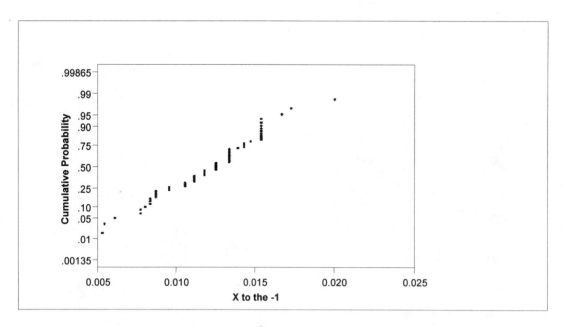

Figure 9.5 Probability Plot of X to the -1 or X^{-1}, the Reciprocal of X

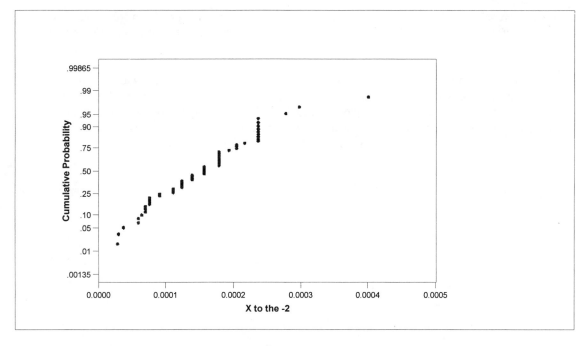

Figure 9.6 Probability Plot of X to the -2 or X^{-2}

To keep a control chart on individual measurements of surgery times for ongoing process control, the procedure suggested here is as follows:

1. Transform the surgery "minutes per procedure," the X values, into their near-normal reciprocals, "procedures per minute."
2. Using the transformed data from step 1, make a temporary I chart on procedures per minute with the control limits based on MRbar for the sole purpose of finding the centerline and limits.
3. Make the inverse transformation of the centerline and control limits of the step 2 temporary I chart, *so these values will again be expressed in the original units*, minutes per procedure.
4. Make an I chart on the initial data in the original units, with the centerline and control limits from step 3.

Keeping the chart on the data in the original units (rather than the transformed units) makes much more sense to the user.

EXAMPLE 9.2

Using the transformed data of X^{-1} from Example 9.1, an I chart is made on the transformed data (Figure 9.7). Note that the control limits are UCL = 0.02042 and LCL = 0.00376, and the centerline is 0.01209 procedures per minute. Taking the inverse transformation, 1 over those values, yields 1/0.02042 = 48.97 and 1/0.00376 = 265.96 minutes for the control limits and 1/0.01209 = 82.71 minutes per procedure for the centerline.

Because a negative power was involved, 1/UCL(transformed data) = LCL(in original units), so LCL = 48.97. Similarly, 1/LCL(transformed data) = UCL(in original units), so UCL = 265.96. Note how these values compare well with the probability plot of Case Study 6.1, Figure 4. The original data are plotted against these limits in Figure 9.8 and are found to be a common-cause system. These control limits would be used for an I chart for ongoing process control. □

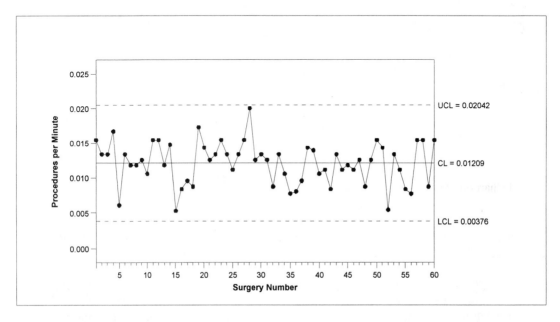

Figure 9.7 I Chart on Transformed Data, Procedures per Minute

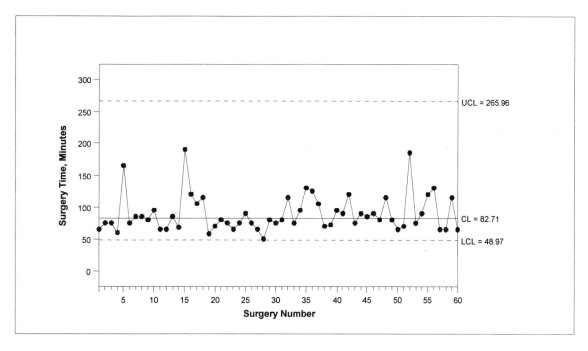

Figure 9.8 I Chart on Original Data, Minutes per Procedure

Rare Events

As a second illustration for the use of transformed data, consider processes with rare events, an important class of problems in health care. Examples of such rare events are both nosocomial and surgical-site infections, other surgical complications, and needlestick accidents, to name a few. If the event rate were high enough, a p or u chart would be used to monitor the event rate. However, when the event rate is very low, the subgroup size required becomes too high for this to be an effective tool. Another approach must be used.

Consider nosocomial infections where the data at hand are the dates of each infection. Queuing theory leads to the expectation that the number of events in a subgroup (e.g., a time period of a month) will have a Poisson distribution. This expectation is met well with real data sets. The c or u chart deals effectively with this problem, provided that the *expected* number of events in each subgroup is sufficiently large. Most statisticians would agree that there is considerable concern when this number falls below four or five. More to the point, all agree that the c and u charts are unsatisfactory when this expected number gets lower than one. But if rare events are to be monitored usefully over time, expected counts below one are the rule, not the exception. Hence, for time-ordered analyses, interarrival intervals must be used. The interarrival interval in this example would be the date of an infection minus the date of the previous infection.

Again falling back on queuing theory, the interarrival intervals are expected to have an exponential distribution. This expectation is also met quite well with real data sets. Without a transformation, the usual

SPC charts (c, u, np, p, *X*, Xbar, and *s*) will *not* work satisfactorily with a distribution that is skewed this severely. A fourth root transformation has often worked well with such data sets.

Benneyan [1998] advocates a new control chart based upon the exponential distribution. The theory is sound, provided that the interarrival times happen to be distributed exactly as predicted, but inventing an entirely new control chart for a single special application and carrying it in our inventory is not inviting. Also, a whole host of new questions would arise as to how robust this proposed chart would be to departures from the postulated theoretical distribution. Cleveland [1993] suggests tying transformations using different powers of the data, selecting a power that yields near-normal data.

EXAMPLE 9.3

Nosocomial methicillin resistant staphylococcus aureus (MRSA) infections have been tracked for 13 years. (See Table 1 in Case Study 9.2 and data set ZMRSA.) The day on which the infection occurred is recorded, so by subtraction, the number of days between infections is calculated. Let day 0 be the day of an infection, say, infection number 0. Then the first infection, infection number 1, occurred on day 16, infection number 2 on day 211, and so on. Therefore, the interarrival times are

for Infection 1: 16 - 0 = 16
for Infection 2: 211 - 16 = 195, and so on.

As seen in Case Study 9.2, a fourth root transformation works well with this particular data set. □

Chapter Summary

- The near-normal assumption is not always met (as seen from the probability plot), particularly for time periods such as surgery times.

- A classic example is "interarrival" intervals of rare events.

- A common way of handling such data is to transform the data to make it near-normal.

- Try a power transformation to transform the data.

- After transformation, check for near-normal again with the probability plot. If still not near-normal, try another transformation and repeat.

- For ongoing process control, make an I control chart on the transformed data. Take the inverse transformation of the control limits and centerline to bring them back to the original limits. Make the I chart with standard given of these values in their original units.

Problems

1.   The data file ZWAIT gives 20 cases for each of five variables: Wait (patient waiting time in minutes), square root of Wait, fourth root of Wait, natural logarithm of Wait, and reciprocal of Wait. Which variables, if any, are satisfactorily near-normal for the use of variables control charts?
8 16 9 10 16 5 4 3 7 13 7 7 11 14 3 20 3 5 22 4

2.   The data file ZSURGERY gives 20 surgery times for surgeon C, repeated here. Try the transformations of square root, fourth root, natural logarithm, and reciprocal of the surgery times. Which, if any, yield satisfactorily near-normal data?
50 60 60 105 52 105 160 83 85 80 95 90 90 195 57 70 105 100 80 60

Case Study 9.1   Surgery Times: Transformations

Background

Recall from the earlier case study on lap chole surgery times after Chapter 6 that scheduling of the operating room has been a problem, and it has been determined that a contributing factor is the large variation in the surgery time for laparoscopic cholecystectomies (lap choles). Several factors of this problem were studied. Total times associated with the surgery were broken into setup time, pre-surgery time, and surgery time. The one considered here is the surgery time. (Refer to Examples 9.1 and 9.2.)

The Data

(Data are in file ZSURG2.) It was found in the case study, through the use of probability plots and histograms, that the surgery times (minutes per procedure) of each surgeon were severely skewed to the right. The Xbar and s chart was used to compare the surgery times of the surgeons. The Xbar and the s charts were "in phase," another indication that the data are severely skewed to the right. Both the s chart and the Xbar chart showed evidence of special-cause variation. Technically, if the s chart is out of control, the limits on the Xbar chart are suspect.

In this chapter it was found that a transformation of the data for surgeon A (taking the reciprocal of the data) made the data near-normal. The data for surgeon E are shown in Table 1.

Case Study 9.1 Table 1 Surgery Times and Transformed Data for Surgeon E

Surgery Number	Surgery Time (minutes per procedure)	Reciprocal (procedures per minute)
1	250	0.004000
2	100	0.010000
3	75	0.013333
4	75	0.013333
5	70	0.014286
6	155	0.006452
7	65	0.015385
8	60	0.016667
9	75	0.013333
10	70	0.014286
11	88	0.011364
12	90	0.011111
13	115	0.008696
14	90	0.011111
15	75	0.013333
16	75	0.013333
17	120	0.008333
18	125	0.008000
19	85	0.011765
20	80	0.012500
21	105	0.009524
22	95	0.010526
23	80	0.012500
24	85	0.011765
25	150	0.006667
26	100	0.010000
27	85	0.011765
28	75	0.013333
29	80	0.012500
30	80	0.012500
31	160	0.006250
32	85	0.011765
33	110	0.009091
34	100	0.010000
35	120	0.008333
36	210	0.004762
37	85	0.011765
38	90	0.011111
39	270	0.003704

Analysis, Results, and Interpretation

The reciprocal transformation was used and worked well on all six surgeons that had more than 24 procedures. For example, the 39 surgery times from surgeon E are shown in Figures 1, 2, and 3. (Fewer than 24 procedures were considered to be too little data to analyze.) The transformed data are shown in Table 1 and Figures 4, 5, and 6.

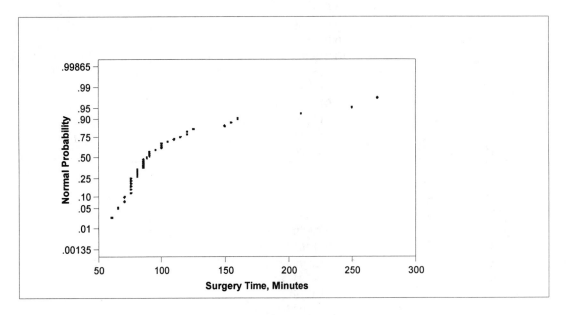

Case Study 9.1 Figure 1 Probability Plot of Lap Chole Surgery Times for Surgeon E

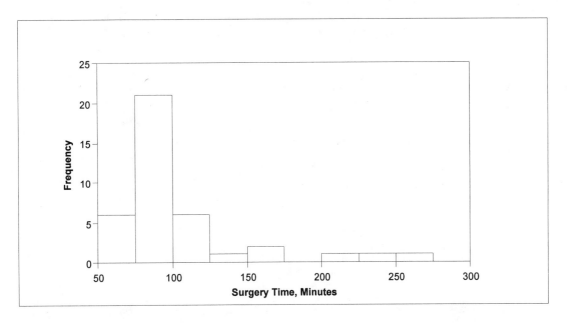

Case Study 9.1 Figure 2 Histogram of Lap Chole Surgery Times for Surgeon E

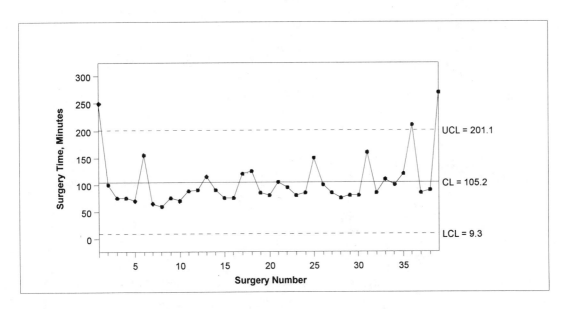

Case Study 9.1 Figure 3 I Chart of Lap Chole Surgery Times for Surgeon E

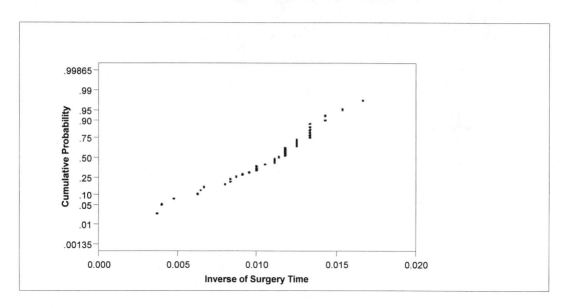

Case Study 9.1 Figure 4 Probability Plot of Inverse of Lap Chole Surgery Times for Surgeon E

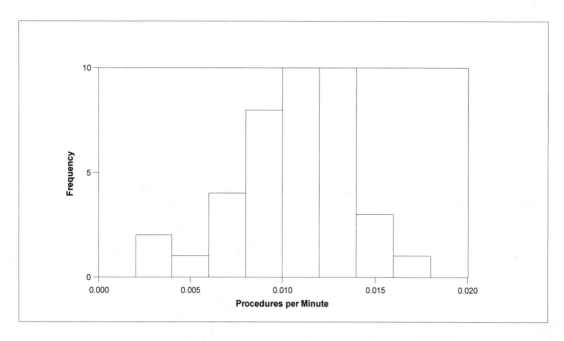

Case Study 9.1 Figure 5 Histogram of Procedures per Minute, the Reciprocal of Minutes per Procedure for Surgeon E

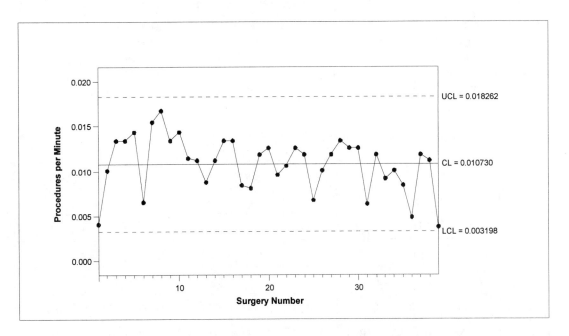

Case Study 9.1 Figure 6 I Chart of Procedures per Minute, the Reciprocal of Minutes per Procedure for Surgeon E

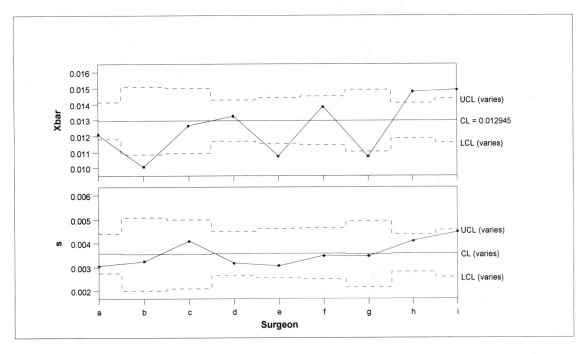

Case Study 9.1 Figure 7 Xbar and s Chart on Procedures per Minute (the Transformed Data) by Surgeon, 2.5-Sigma Limits

Recall that in Case Study 6.1, Figure 5, an attempt was made to compare (profile) the surgeons with an Xbar and s chart. That Xbar and s chart was unsatisfactory because the Xbar chart and the s chart were in phase and because the s chart was out of control. In the Xbar and s chart on the transformed data (Figure 7 with relevant information in Table 2) both of these deficiencies have been remedied. Note that in this case the same surgeons were out of control before and after the transformation.

Case Study 9.1 Table 2 Statistics on Transformed Data by Surgeon, 2.5-Sigma Limits

	Surgeon								
	A	B	C	D	E	F	G	H	I
Xbar	.01209	.01007	.01263	.01322	.01073	.01381	.01070	.01475	.01487
s	.00304	.00325	.00408	.00316	.00306	.00347	.00346	.00408	.004447
n	60	18	20	48	39	35	22	66	43
UCL(Xbar)	.01411	.01506	.01495	.01424	.01438	.01446	.01486	.01486	.01432
LCL(Xbar)	.01179	.01083	.01094	.01165	.01151	.01143	.01103	.01184	.01157
UCL(s)	.00440	.00507	.00500	.00450	.00460	.00465	.00493	.00437	.00455
sBar	.00358	.00354	.00355	.00358	.00357	.00357	.00355	.00358	.00357
LCL(s)	.00275	.00201	.00210	.00265	.00254	.00248	.00217	.00279	.0026

Lessons Learned

Measurement data that are severely skewed need to be transformed to near-normal so that the assumptions are met for variables control chart analysis. For instance, data need to be near-normal to make an I chart. On an Xbar and s chart, the s chart needs to be in control to validate the Xbar limits. Also, the data should not be so severely skewed that the Xbar chart and the s chart are in phase.

Managerial Considerations

As mentioned earlier, since there is a large difference between surgeons, it may be helpful to have the surgeons discuss among themselves their approach to the procedure. In this way they may learn from each other and perhaps standardize some aspects of the procedure.

Problems

1.   Using ZSURG2, what transformation best suits the data for surgeon D? Surgeon H?

Case Study 9.2 Decreasing the Nosocomial MRSA Infection Rate

Background

Refer to Example 9.3. Records from this hospital are available for about the last 13 years giving the dates of occurrence of nosocomial methicillin resistant staphylococcus aureus (MRSA) infections. Ongoing efforts have been made over this time period to decrease the infection rate, that is, to *increase* the intervals of the MRSAs. In order to determine what type of action to take at this time for process improvement, it is necessary to determine whether past efforts have resulted in special-cause variation over time— improvement that cannot be attributed simply to common-cause variation. A method is also desired to monitor the process over time for ongoing process control.

The Data

(Data are in file ZMRSA.) The 4,777 consecutive days of records have been serially numbered with day number 1 assigned to the first day after the first recorded MRSA. In Table 1, Obs is the observation (MRSA) number; DayNumber is the serial number of the day on which each of the MRSAs occurred. The variable named Interval is the interval of the successive infections, measured in days. (Hence, DayNumber is the cumulative sum of the intervals.) The table shows that there were a total of 66 MRSAs in the 4,777 days. The variable phase in Table 1 divides the data into two time periods of 33 infections each. Note that phase 1 covered 1,447 days and phase 2 contains the remainder, 4,777 - 1,447 = 3,330 days.

Case Study 9.2 Table 1 MRSA Data

Obs	FourthRoot	Interval	DayNumber	Phase
1	2.00	16	16	1
2	3.74	195	211	1
3	2.93	74	285	1
4	2.60	46	331	1
5	1.93	14	345	1
6	1.57	6	351	1
7	2.36	31	382	1
8	1.50	5	387	1
9	1.82	11	398	1
10	2.40	33	431	1
11	2.17	22	453	1
12	2.65	49	502	1
13	1.32	3	505	1

14	1.32	3	508	1
15	2.72	55	563	1
16	1.57	6	569	1
17	2.41	34	603	1
18	1.86	12	615	1
19	2.99	80	695	1
20	2.62	47	742	1
21	3.61	169	911	1
22	1.68	8	919	1
23	3.03	84	1003	1
24	3.16	100	1103	1
25	1.57	6	1109	1
26	1.73	9	1118	1
27	1.68	8	1126	1
28	2.36	31	1157	1
29	2.40	33	1190	1
30	2.66	50	1240	1
31	3.25	112	1352	1
32	2.74	56	1408	1
33	2.50	39	1447	1
34	3.84	217	1664	2
35	1.73	9	1673	2
36	1.57	6	1679	2
37	2.77	59	1738	2
38	1.32	3	1741	2
39	3.34	124	1865	2
40	1.97	15	1880	2
41	3.99	254	2134	2
42	3.67	182	2316	2
43	1.68	8	2324	2
44	3.49	148	2472	2
45	2.88	69	2541	2
46	3.05	87	2628	2
47	2.90	71	2699	2
48	1.86	12	2711	2
49	2.51	40	2751	2
50	2.11	20	2771	2
51	1.32	3	2774	2
52	4.58	440	3214	2
53	2.72	55	3269	2
54	2.24	25	3294	2
55	3.58	165	3459	2
56	2.85	66	3525	2
57	3.09	91	3616	2
58	2.84	65	3681	2
59	1.82	11	3692	2

60	1.57	6	3698	2
61	2.85	66	3764	2
62	4.27	333	4097	2
63	2.96	77	4174	2
64	3.19	103	4277	2
65	2.87	68	4345	2
66	4.56	432	4777	2

Analysis, Results, and Interpretation

Figure 1 is a run chart on the time-ordered intervals between infections measured in days. There appears to be a trend upward through the entire time period, which would be evidence of improvement (increase in length of intervals) if confirmed.

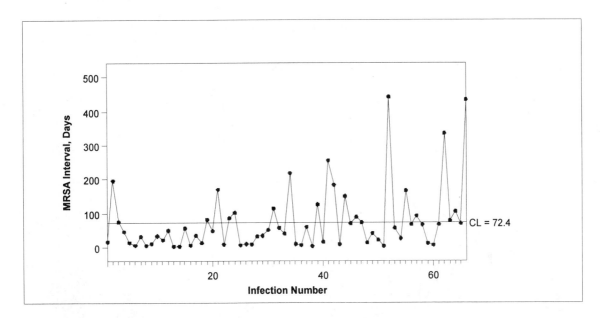

Case Study 9.2 Figure 1 Run Chart of MRSA Infection Interval Measured in Days

The infection intervals will be analyzed as variables data, which requires that their distribution be near-normal. However, it is anticipated that the intervals will be badly skewed to the right so that a transformation will be needed. This is confirmed in the histogram and probability plot, Figures 2 and 3.

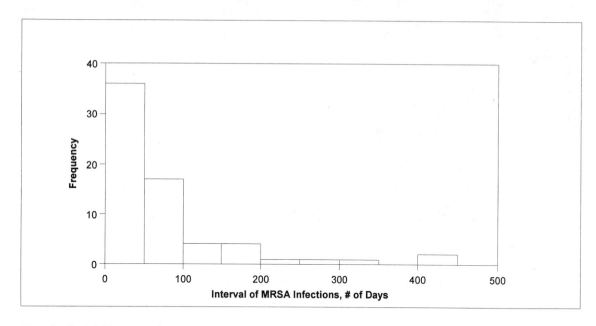

Case Study 9.2 Figure 2 Histogram of MRSA Infection Interval Measured in Days

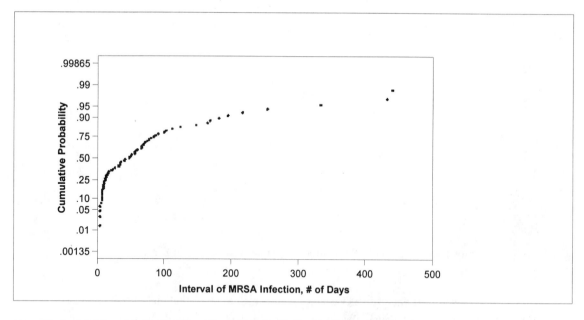

Case Study 9.2 Figure 3 Probability Plot of MRSA Infection Interval Measured in Days

A fourth root transformation will often transform intervals such as these to near-normal data. The histogram and probability plots, Figures 4 and 5, show the transformed distribution to be satisfactorily near-normal for the making of control charts.

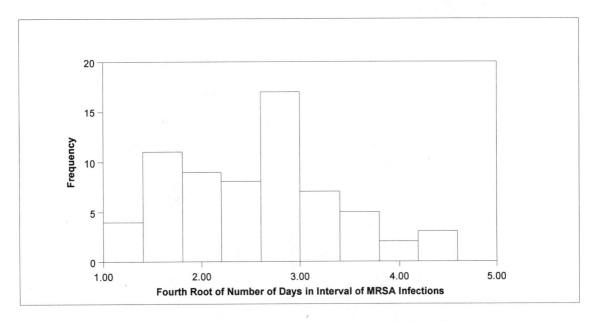

Case Study 9.2 Figure 4 Histogram of Fourth Root of Number of Days in Interval

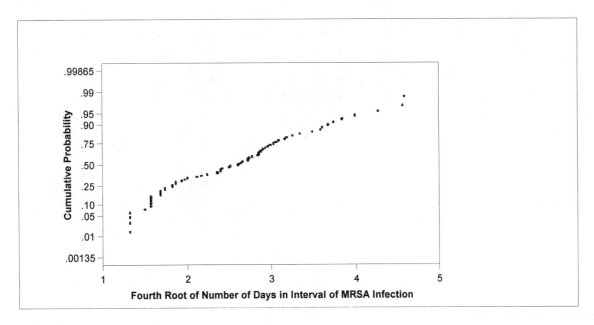

Case Study 9.2 Figure 5 Probability Plot of Fourth Root of Interval

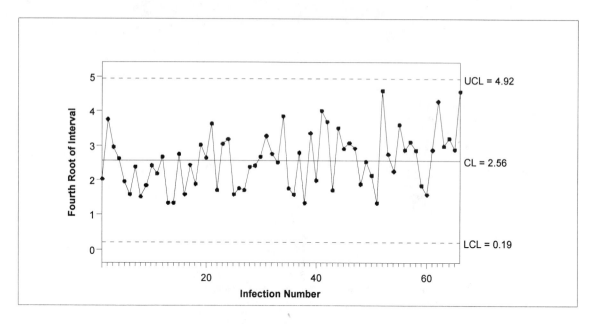

Case Study 9.2 Figure 6 I Chart on Fourth Root of Interval

The fourth root transformation used in Figure 6 is unnecessarily complicated to use for ongoing process control, but there is an easy way around this. The centerline and control limits from Figure 6 may be used to find values for those quantities expressed in days. For example, the centerline in Figure 6 is 2.56. If this is raised to the fourth power (i.e., squared and then squared again), the result is 42.9 days. That is the correct centerline to be used for the I chart on intervals in days. The complete chart is shown in Figure 7.

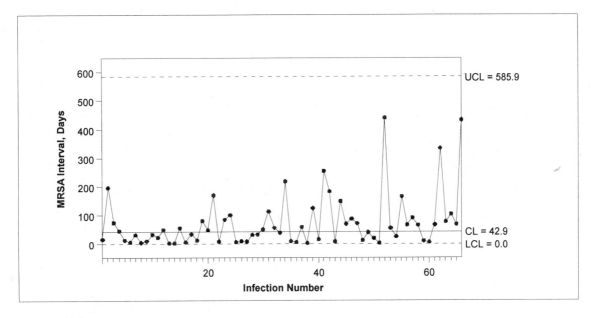

Case Study 9.2. Figure 7 I Chart on Interval in Days with Centerline and Limits Derived from Fourth Root I Chart

The I chart in Figure 7 is correct but not very useful. It lacks the necessary power to demonstrate evidence of historical improvement and is hardly satisfactory for ongoing process control. There can be no point above the upper control limit, evidence of process improvement, until an interval of 586 days occurs between MRSA infections. Perhaps the control chart should not be faulted for this; the process being "too good" carries the price of difficulty in detecting improvement. Since the lower control limit is zero, a deterioration of the process can only be shown by a long run below the centerline.

The power of the control chart can be increased by using subgroups of data—the larger the subgroup the more analytical power. By way of example, the data set containing 66 MRSA infections has been broken into two time "phases," each with 33 infections, as shown in Table 1.

Figure 8 is an Xbar and s chart on the fourth roots of the intervals using just two subgroups, one for each phase. The s chart is in control, so the Xbar chart is valid. The Xbar chart shows special-cause variation; the time intervals between infections were so large in the second time period that the improvement could not reasonably be attributed to random chance.

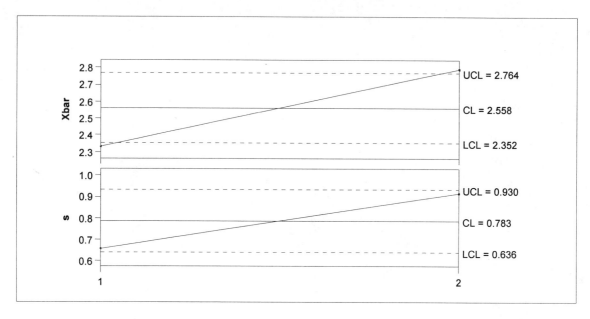

Case Study 9.2 Figure 8 Xbar and s Chart on Fourth Root of Infection Interval, Phase 1 versus Phase 2, 1.5-Sigma Limits

A simpler way to make the comparison between the two phases does not require transformation of the original data. The comparison of the two phases may also be made using a u chart, where u is the infection rate (infections per day). From the data section, it is seen that the first phase had 33 infections in 1,447 days and the second phase had 33 infections in 3,330 days. The u chart is shown in Figure 9. The results are consistent with what has already been seen; the improvement in phase 2 was special-cause variation, not random chance. In this case study, the u chart provided the simplest method of analysis to show that there had been special-cause improvement during the 13-year period.

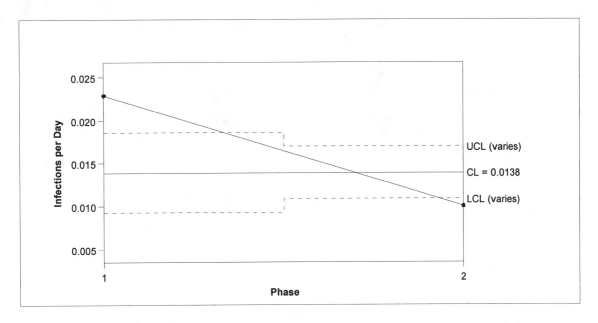

Case Study 9.2 Figure 9 u Chart on Infections per Day, Phase 1 versus Phase 2, 1.5-Sigma Limits

Lessons Learned

A powerful method of detecting a trend in time-ordered data is to divide the data in half and analyze it with two subgroups. A data set for rare events often lends itself to the use of an Xbar and s chart on the fourth root of the intervals and/or the use of a u chart on the infection rate. Using both approaches provides a check on the results.

As the nosocomial MRSA infections become more rare, providing a useful monitoring chart becomes more difficult. One approach would be to try a u chart with subgroups of a half year or a year.

Management Considerations

Management should have comparatively little trouble obtaining a useful analysis of historical data following the methods used in this case study. However, finding a useful approach to ongoing monitoring may be a real challenge. The more rare the event, the more difficult it is going to be to find a useful method of following it over time.

Problems

1.   Does using an Xbar and s chart on the transformed intervals with 22 subgroups of size 3 provide sufficient power to detect the process improvement?

2.   Does using an Xbar and s chart on the transformed intervals with 11 subgroups of size 6 provide sufficient power to detect the process improvement?

3.   Does using an Xbar and s chart on the transformed intervals with 6 subgroups of size 11 provide sufficient power to detect the process improvement?

4.   Does using an Xbar and s chart on the transformed intervals with 3 subgroups of size 22 provide sufficient power to detect the process improvement?

Case Study 9.3 Reducing Surgical Site Infections

Background

The hospital has maintained records on surgical site infections (SSI) for the past 16 months, with all types of surgical procedures combined. During the tenth month, an improvement program was introduced that became effective with a new pathway at the end of that month. Now the hospital needs to determine whether there has been a reduction in the SSI rate since the date of the new pathway.

The Data

(Data are in file ZSSI.) Table 1 shows the data for surgical site infections for the past 16 months. "Day zero" was the date of a procedure that became infected and is not included in the Table 1 data. The new pathway date was after the 53rd infection, which occurred at 302 cumulative days as noted in the variable Phase.

Case Study 9.3 Table 1 Surgical Site Infection Data

A = Infection Number
B = Days Cumulative
C = Interval in Days
D = Fourth Root of Interval
E = Phase

A	B	C	D	E
1	4	4	1.41	1
2	12	8	1.68	1
3	15	3	1.32	1
4	17	2	1.19	1
5	22	5	1.50	1
6	29	7	1.63	1
7	35	6	1.57	1
8	46	11	1.82	1
9	48	2	1.19	1
10	55	7	1.63	1
11	55	0	0.00	1
12	56	1	1.00	1
13	69	13	1.90	1
14	91	22	2.17	1

15	106	15	1.97	1
16	108	2	1.19	1
17	119	11	1.82	1
18	122	3	1.32	1
19	127	5	1.50	1
20	129	2	1.19	1
21	133	4	1.41	1
22	135	2	1.19	1
23	138	3	1.32	1
24	140	2	1.19	1
25	142	2	1.19	1
26	150	8	1.68	1
27	151	1	1.00	1
28	157	6	1.57	1
29	161	4	1.41	1
30	162	1	1.00	1
31	162	0	0.00	1
32	173	11	1.82	1
33	173	0	0.00	1
34	177	4	1.41	1
35	183	6	1.57	1
36	202	19	2.09	1
37	203	1	1.00	1
38	219	16	2.00	1
39	219	0	0.00	1
40	221	2	1.19	1
41	241	20	2.11	1
42	242	1	1.00	1
43	249	7	1.63	1
44	253	4	1.41	1
45	255	2	1.19	1
46	260	5	1.50	1
47	266	6	1.57	1
48	268	2	1.19	1
49	269	1	1.00	1
50	279	10	1.78	1
51	282	3	1.32	1
52	299	17	2.03	1
53	302	3	1.32	1
54	312	10	1.78	2
55	342	30	2.34	2
56	351	9	1.73	2
57	356	5	1.50	2
58	357	1	1.00	2
59	360	3	1.32	2
60	385	25	2.24	2

61	393	8	1.68	2
62	419	26	2.26	2
63	453	34	2.41	2
64	460	7	1.63	2
65	465	5	1.50	2
66	467	2	1.19	2
67	471	4	1.41	2
68	492	21	2.14	2

Analysis, Results, and Interpretation

There is no information on SSIs for different types of procedures and the mix of surgical procedures is thought to have been relatively stable, so the best solution until better data is available might be the calculation of a single rate for all SSIs combined. A better approach might also be to do nothing until it can be done well.

From Table 1 it may be seen that the time era up through the intervention included 53 infections in 302 days and the time era after the intervention included 15 infections in 190 days. The u chart (Figure 1) uses 1.5-sigma limits and shows that the improvement in the SSI rate was not just random chance but a special cause of variation.

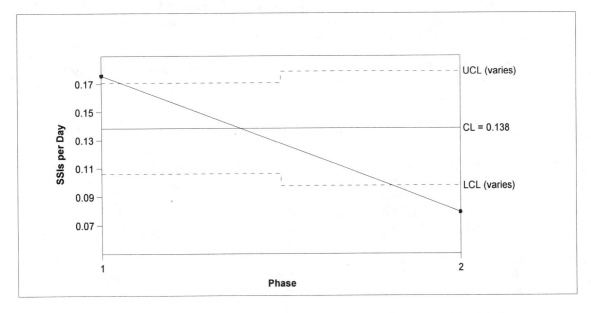

Case Study 9.3. Figure 1 u Chart for SSI Rate before and after New Pathway, 1.5-Sigma Limits

Lessons Learned

Since some surgical procedures are much more prone to SSI than others, the data collection should be done by procedure or by families of procedures with similar risk of infection. The data can always be pooled later, but if pooled initially, they cannot be readily separated.

Aside for the concern about pooled data, for a case such as this where the objective is simply to compare "before" and "after," there is no need for an analysis more complicated than that done here. Other analysis possibilities are explored in the Problems.

Management Considerations

It appears that, with the pooled data, there has been an overall improvement. However, management should be quick to ask if there are better ways for the data to be collected and to make the resources available for that purpose. Indiscriminate pooling as was done here is very dangerous. See Appendix 2 for more information.

Problems

1.   Verify with a histogram and a probability plot that the distribution of the 68 infection intervals in data file ZSSI is severely skewed to the right as expected.

2.   Look at the distribution of the fourth root transformation of the same 68 intervals in ZSSI with a histogram and probability plot to see if the distribution is now near-normal. Pay particular attention to the difficulty caused by the four zero values.

3.   Verify that in data file ZSSI2 the four zero values for the intervals in days have been replaced with values of 0.5 days. What does a value of 0.0 mean? What does a value of 0.5 mean? Verify that the fourth roots of the 0.5 values are correct.

Use data file ZSSI2 for the following problems:

4.   Verify with a histogram and probability plot that the fourth root transformations of the 68 intervals in ZSSI2 are now satisfactorily near-normal.

5.   Using probability plots, verify that the distributions of the fourth roots are satisfactorily near-normal for the 53 infection intervals of phase 1 and for the 15 infection intervals of phase 2.

6.   Make an I chart on the 68 fourth roots. Although it may raise suspicions that the last 15 points (the phase 2 points) show improvement, note that this I chart lacks sufficient power to support these suspicions.

7.   Make an I chart on the 53 fourth roots from phase 1. Is there evidence of special-cause variation?

8.   Make an I chart with standard given on the 15 fourth roots from phase 2 using the results from phase 1 as standard values. Note that there are no points outside of the control limits but that there would be evidence of special-cause variation if the test for two out of three points beyond the same 2-sigma limit were being used.

9.   Could an I chart have been made on phase 1 with standard given using phase 2 for the standard values? What reasons are there to say that the way it was done above is better?

10.   Make an I chart for the 15 fourth roots from phase 2 with no standard given. Might it be useful to use standard values from this chart for ongoing process control?

11.   Make an Xbar and s chart on the fourth roots with one subgroup for each phase. What value of T should be used for the T-sigma limits? Is the Xbar chart validated by the s chart being in control? What is learned from the Xbar chart? How does this compare with what was learned from the u chart in this case study?

12.   What is the best control chart approach for comparing the two time phases? Why?

Case Study 9.4 Surgery Complications with New Cost Savings Pathway

The concept for this case study was developed with the assistance of Dr. Ray Carey of R. G. Carey and Associates, Park Ridge, IL.

Background

A new cost savings pathway has been introduced for bowel surgery, and statistical evidence is needed to show that there has not been any deterioration in outcome quality. An important consideration is to verify that the frequency of surgery complications has not increased.

If the complication rate were high enough, a p chart could be used to monitor the percentage of procedures that were complicated. However, when the complication rate is very low, the number of procedures required to make a subgroup becomes too high for this to be an effective tool. It is then said that a complication is a "rare event" for which another approach must be used.

For rare events, instead of studying the complication rate itself, the intervals between complications are analyzed. For surgery complications, this interval is measured as the number of procedures between complications.

An initial study of the complication intervals had incorrectly concluded that the complication rate had gotten worse after the introduction of the new pathway. This case study shows how and why that wrong conclusion was drawn and how to properly analyze rare events.

The Data

(Data are in file ZCOMPLI.) Bowel surgery procedures are numbered, with number one being the first procedure after a complication. Only the complicated procedures are recorded in Table 1. Complications are assigned a serial number, given in the first column of the table. The first 25 complicated procedures occurred in era 1, as shown in the second column. The procedure number of each complicated procedure is given in column 3. The interval between complications, measured in number of procedures, is given in column 4. An explanation of column 5 will be given in the Analysis section.

Case Study 9.4 Table 1 Surgery Complication Data

SN	Era	Proc	Int	Trans	SN	Era	Proc	Int	Trans
1	1	16	16	2.00	31	2	206	6	1.57
2	1	20	4	1.41	32	2	220	14	1.93
3	1	22	2	1.19	33	2	227	7	1.63
4	1	34	12	1.86	34	2	235	8	1.68
5	1	56	22	2.17	35	2	249	14	1.93
6	1	62	6	1.57	36	2	252	3	1.32
7	1	65	3	1.32	37	2	254	2	1.19
8	1	66	1	1.00	38	2	255	1	1.00
9	1	70	4	1.41	39	2	260	5	1.50
10	1	76	6	1.57	40	2	271	11	1.82
11	1	83	7	1.63	41	2	276	5	1.50
12	1	87	4	1.41	42	2	281	5	1.50
13	1	110	23	2.19	43	2	290	9	1.73
14	1	112	2	1.19	44	2	302	12	1.86
15	1	115	3	1.32	45	2	303	1	1.00
16	1	118	3	1.32	46	2	309	6	1.57
17	1	122	4	1.41	47	2	310	1	1.00
18	1	148	26	2.26	48	2	313	3	1.32
19	1	159	11	1.82	49	2	333	20	2.11
20	1	162	3	1.32	50	2	335	2	1.19
21	1	168	6	1.57	51	2	345	10	1.78
22	1	170	2	1.19	52	2	347	2	1.19
23	1	176	6	1.57	53	2	354	7	1.63
24	1	180	4	1.41	54	2	365	11	1.82
_25_	_1_	_182_	_2_	_1.19_	55	2	368	3	1.32
26	2	190	8	1.68	56	2	371	3	1.32
27	2	193	3	1.32	57	2	376	5	1.50
28	2	194	1	1.00	58	2	378	2	1.19
29	2	196	2	1.19	59	2	381	3	1.32
30	2	200	4	1.41	60	2	383	2	1.19

SN	The complication serial number
Era	Era 1 is before new pathway: era 2 is after
Proc	Procedure number (procedure zero was complicated)
Int	Number of procedures in interval between complications
Trans	Transformed data, fourth root of complication interval

Analysis, Results, and Interpretation

Figure 1 is a run chart for the surgery complication intervals, showing a line of demarcation between eras one and two where the new pathway was introduced after complication number 25. You might have some suspicion that the new pathway was detrimental, that is, followed by shorter intervals between complications. Such concern is appropriate and should lead you to seek statistical evidence rather than just hunches.

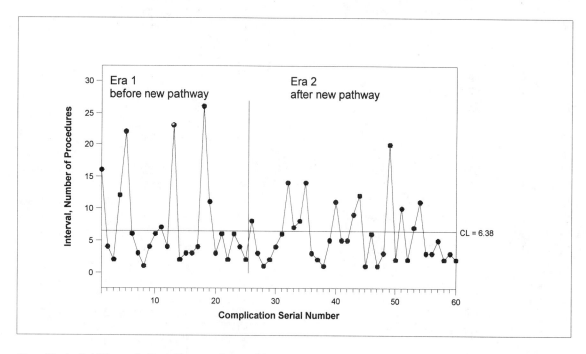

Case Study 9.4 Figure 1 Run Chart on Interval between Complicated Procedures (Expressed in Number of Procedures)

In an ill-conceived attempt to obtain statistical evidence of special-cause variation, the I chart of Figure 2 was made on the complication intervals. It shows three points above the UCL in era 1 and none in era 2. With this chart it incorrectly appears that there is special-cause variation between the two eras; it looks like the process deteriorated after the introduction of the new pathway. This was the *wrong* conclusion; it would have been a valid conclusion if the control chart had been valid—but the control chart was not valid as will now be explained.

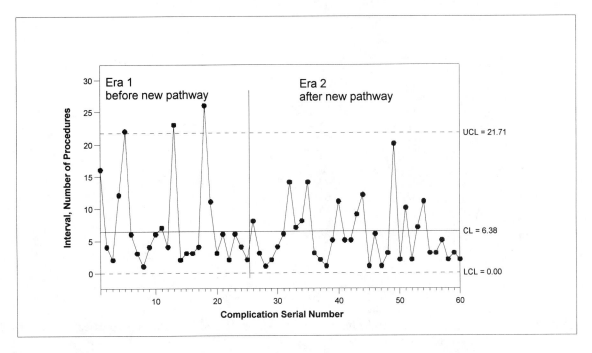

Case Study 9.4 Figure 2 I Chart on Interval between Complicated Procedures (Expressed in Number of
Procedures)

The control chart is valid only if the data distribution is near-normal, but that is not the case. Theory tells us
that these intervals will be severely skewed to the right. The histogram (Figure 3) on the complication
intervals shows that the data are not near-normal but are instead badly skewed to the right, violating the
assumption of near-normality.

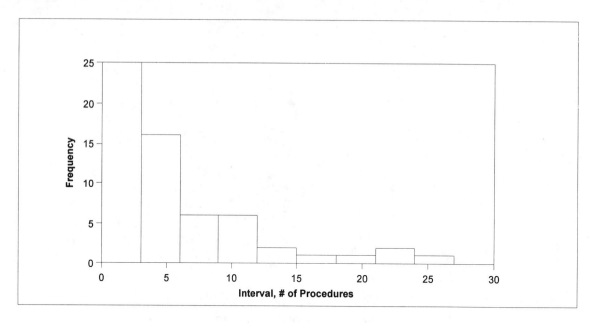

Case Study 9.4 Figure 3 Histogram of the Interval between Complicated Procedures (Expressed in Number of Procedures)

A more discriminating graphical test for normality is the normal probability plot [Shapiro, 1990]. If near-normal, the data on the normal probability plot fall approximately along a straight line. The normal probability plot (Figure 4) does not yield a straight line.

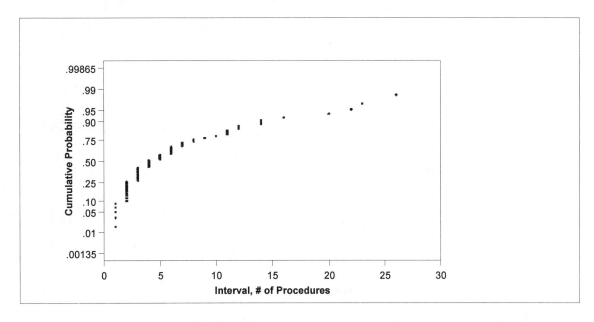

Case Study 9.4 Figure 4 Probability Plot of the Interval between Complicated Procedures (Expressed in Number of Procedures)

If an I chart is incorrectly made, using data that are severely skewed to the right, you cannot tell whether points above the UCL are due to the skewness or due to special-cause variation over time. When data are severely skewed to the right, the options are to use specialized control charts, such as those suggested by Benneyan [1998], or to transform the data to make it near-normal. Cleveland [1993] suggests trying transformations using different powers of the data, selecting a power that yields near-normal data. This approach provides the simplicity of not needing specialized control charts. For distributions such as these for rare events, the fourth root of the original data often yields near-normal data.

The column titled "Trans" in the data file is the fourth root of the complication interval. For instance, the first two procedures with complications were procedures numbered 16 and 20, which had intervals of 16 and 4 procedures, respectively. The fourth roots of 16 and of 4 are 2 and 1.41, respectively. The histogram and probability plots (Figures 5 and 6) verify that the fourth root transformation produces data that are near-normal. (Compare these to Figures 3 and 4.)

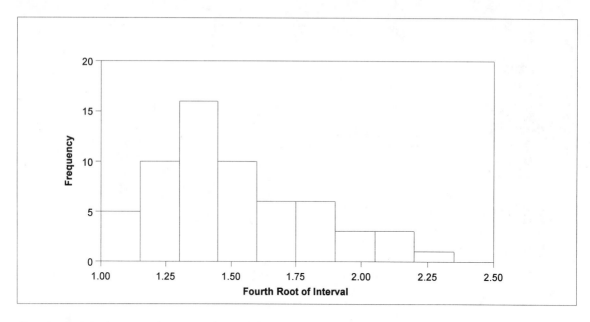

Case Study 9.4 Figure 5 Histogram of Fourth Root of Complication Interval

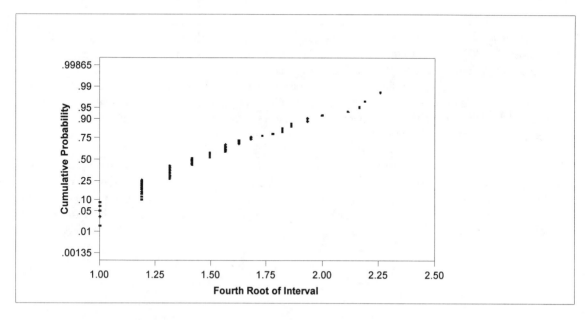

Case Study 9.4 Figure 6 Probability Plot of Fourth Root of Complication Interval

The I chart in Figure 7 shows the process to be in control, finding no increase in surgery complications after the introduction of the new pathway. For this example, an I chart would be useful for ongoing process control, but would be much easier to understand if it were made on the original complication intervals (in number of procedures) rather than on the fourth root. To accomplish this, the centerline and control limits from Figure 7 are raised to the fourth power and used as illustrated in Figure 8. Compare this correct control chart with the incorrect one in Figure 2. This final I chart shows the intervals in number of procedures. Note that an interval greater than 36 procedures would be evidence of a process improvement.

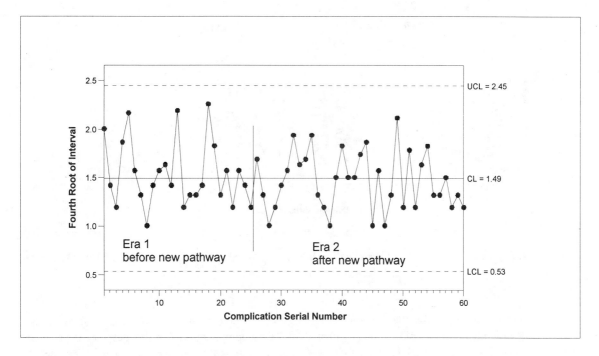

Case Study 9.4 Figure 7 I Chart on Fourth Root of Complication Interval

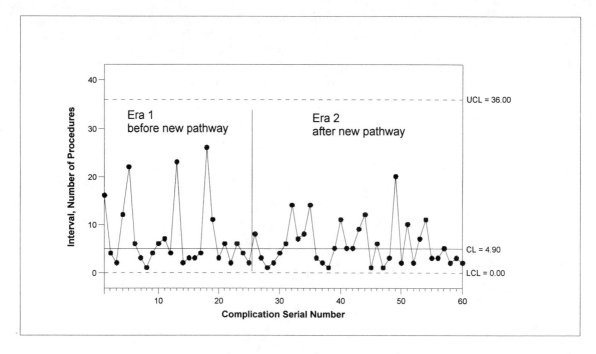

Case Study 9.4 Figure 8 I Chart on Complication Interval Using Standard Given; Standard Values Adapted from Fourth Root I Chart

The I chart shown in Figure 8 is the best approach that can be made for ongoing process control, allowing the complication interval to be plotted and evaluated after each complication. For this purpose, the standard values from era 2 should be used to analyze the new data with standard given, as that new data becomes available.

Other possibilities exist for a retrospective analysis comparing the two eras. With an I chart on the transformed intervals, use standard values from era 2 to analyze era 1 with standard given. It is possible to obtain more power by a retrospective analysis using larger subgroups. This could be done here by using just two subgroups, one before the new pathway and one after. The data could then be analyzed with a p chart on the complication rate, an Xbar and s chart on the transformed complication intervals, or both. (See problems 1 through 3.)

Lessons Learned

When time intervals are involved, the data should always be checked with a probability plot to determine whether it will pass as near-normal for control chart analysis. This is particularly true for the intervals between rare events. When the data are not near-normal, an appropriate transformation must be used.

Management Considerations

The need to analyze rare events does not arise in most manufacturing or service organizations. In health care it is a way of life. A considerable waste of money would have been incurred by the hospital if the initial incorrect analysis had not been replaced with one that was correct. Management decisions must be made based on data, but a superficial knowledge of statistical process control methods cannot provide an adequate interpretation of that data.

Problems

1. With an I chart on the transformed intervals, use standard values from era 2 to analyze era 1 with standard given. Is there evidence of special-cause variation between the two eras?

2. Make a p chart on the surgery complication rate using one subgroup for each era. Is there evidence of special-cause variation between the two eras?

3. Make an Xbar and s chart on the transformed intervals between complications using one subgroup for each era. Is there evidence of special-cause variation between the two eras?

4. For era 2 only, is the I chart on the transformed data in control?

Chapter 10 Guidelines for Making Control Charts Useful

Choice of Control Chart

When making a control chart, the decision tree in Figure 10.1 provides a convenient guide to the choice of a control chart for a particular application.

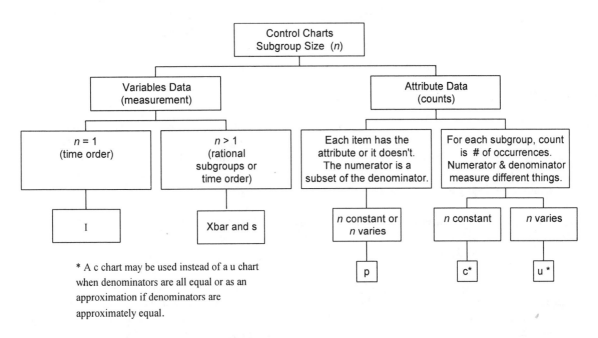

Figure 10.1 Decision Tree for Control Chart Choice

The first decision on the tree is whether the data are variables data or attribute data. Variables data are usually the result of measurements made upon quantities that are continuous by nature, such as length or time. Attribute data are the result of counting and are of two types:

1. The c chart and the u chart assume a *Poisson distribution* and plot the count of the number of occurrences of a characteristic in a unit. A classic example is the count of the number of raisins in a slice of raisin bread. A typical healthcare application is the study of the monthly count of patient falls. A *count* does not exist without an *area of opportunity* (or "exposure") in which this count may occur. This area of opportunity, or *subgroup size*, is expressed in convenient *units*. For a monthly count of patient falls, the area of

opportunity might be expressed in units of 100 patient days each. If the number of patient days each month (i.e., the area of opportunity or subgroup size) is essentially constant, the c chart may be used. If the subgroup sizes vary significantly (e.g., the number of patients differs significantly from month to month), a u chart is required rather than the simpler c chart. Note that there can be more than 100 falls in 100 patient days; theoretically, there is no upper limit on the count.

2.   The p chart assumes a *binomial distribution* where, for a group of like events, a *special outcome* is either present or absent. The value *p* for the p chart is the proportion of the events where the special outcome occurs (which may be expressed as a percentage). To illustrate with the monthly incorrect medication rate, let the special outcome be an incorrect medication dose. The p value for a month is the number of incorrect doses divided by the *total* number of doses for the month. For a month, the number of incorrect doses plus the number of correct doses equals the total number of doses, which is the subgroup size. The numerator is a subset of the denominator; the number of incorrect doses cannot exceed the total number of doses.

Miscellaneous Gems

These gems were collected from a variety of teachers (which included many students) over the years. Each control chart assumes a particular distribution, such as near-normal for the variables charts. Although the control charts are quite forgiving in this regard, variables distributions should be checked for being too skewed using a probability plot. For attribute data, protection against being too severely skewed is given by the sampling and subgrouping scheme.

It is assumed that consecutive observations are independent. This assumption is best acknowledged and then ignored because it is seldom met in real process data. If it could not be ignored, there would be very few control charts except those made on tables of random numbers.

The most important consideration regarding the data is the sampling scheme. Study the process; look for places where special-cause variation is likely to enter, and sample the process there. Be careful to sample from all conditions that exist in the process—and to label the numerical data that are acquired accordingly. The numbers that are collected are only half of the data. The other half is the set of circumstances under which each value is obtained.

Subgroups should be as homogeneous as possible. This means that time-ordered values for an I chart should be as close together in time as possible, with an interruption noted wherever necessary. (See "The Interrupted I Chart" in Chapter 4.)

Wherever possible, the time order of the data should be maintained and the analysis made in time order. With any time-ordered control chart that uses subgroups, all process changes should come between subgroups, not within one subgroup. If you opt to use an Xbar and s chart with small subgroups for time-ordered analysis, a check should first be made for within-subgroup stratification.

Annotate the charts (including run charts). Comment wherever it may be helpful, particularly regarding what was found when suspicious results were investigated. There are two kinds of notes: those noting what was discovered *relating to one point* and those telling about process changes that occurred *between points*. Draw an arrow with the note showing exactly to which point or to which line segment (or gap between points) it applies.

When an Xbar and s chart is made to compare rational subgroups, an I chart should be made for each subgroup on its elements in time order. When an I chart is made on a single stream, there is often an opportunity to compare that stream with others, using an Xbar and s chart. This same concept applies to attribute data; for example, when using a p chart to compare rational subgroups, you should look within each subgroup to see that the occurrences of interest are randomly dispersed within that subgroup over time—not all bunched up in one short time period.

Each chart should be "self-contained" to the extent possible. Start by giving the chart a good title. State, perhaps in the title or subtitle, what kind of chart it is (e.g., a p chart). The user expects the chart to have 3-sigma limits. If that is not the case, the chart should state the width of the limits that are used. The user should also be told on the chart or in an accompanying table the values of the centerline, control limits, plotted points, and subgroup sizes. State on the chart whenever it is made with standard given and tell the source of the standard values. The raw data should be available to the user, either in hard copy or electronically.

For attribute charts, verify that the subgroup sizes are large enough for the results to be believed.

For charts with varying subgroup sizes, the control limits step up and down, being wider for the smaller subgroups since less certainty exists there. A further complication on the s chart is that there will also be steps in the sBar line.

Check the software! Does it give the right answers for an interrupted I chart and for an Xbar and s chart with varying *n*?

With ongoing process control, beware of overcorrecting. Taking action on the process when it should be left alone will increase the total variation of the process.

Problems

For problems 1 – 10 (repeated from Chapter 1), tell which chart is most applicable.

1.   Length of stay subgrouped by diagnosis-related group (DRG).

2.   Cost per case for each case in order of time.

3.   Accounts receivable: days since billing subgrouped by payer.

4.   Percentage of primary cesarean sections (subgrouped by physician, payer, month).

5. Number of cardiac rehab patient visits per week.

6. For each patient, the time is recorded from the time of the decision to admit the patient to the time of transfer to the unit.

7. Time from order entry to result.

8. Recorded each day: the number of reports that are more than three days old.

9. Percentage of medication errors.

10. Number of critical care unit (CCU) infections per patient day subgrouped by month.

Case Study 10.1 Benchmarking Time to Extubation after CABG Surgery

The concept for this case study was developed with the assistance of Dr. Ray Carey of R. G. Carey and Associates, Park Ridge, IL.

Background

In early April, the five anesthesiologists met to discuss best methods for minimizing the time to extubation after uncomplicated coronary artery bypass graft (CABG) surgery. They decided to meet again after obtaining the historical data on length of intubation (LOI) for the first three months of the year.

The first part of the case study analyzes the baseline three-month period with the objective of process improvement.

The Data

(Data are in files noted.) The data recorded for each procedure were a time-ordered serial number (SN 1 through 61), the month, anesthesiologist identification number (ANES), and the LOI to the nearest hour. These LOI data are then measurement data, so I and Xbar and s charts are appropriate on these measurements. When extubation was done in the operating room (OR), the procedure was assigned an LOI of 0 hours. It was agreed that complicated procedures would be defined as those where LOI exceeded 24 hours.

Baseline Three Months: Data

There had been 61 CABG procedures during months 1 through 3, which were reviewed in month 4 by the five anesthesiologists. The initial data file, ZLT3, has been sorted first on LOI, then on ANES, as shown in Table 1.

Baseline Three Months: Analysis, Results, and Interpretation

By scanning Table 1 the following can be seen:

1. The cutoff LOI time of 24 hours makes sense; there is a very large gap between the highest uncomplicated time of 20 hours and the lowest complicated time of 84 hours.
2. ANES #1 had all four of the cases where the extubation was done in the OR.

3.　　　All except three of the lowest 15 LOIs were from ANES #1.

Case Study 10.1 Table 1 Baseline Three Months: LOI Times, Hours

SN	MONTH	ANES	LOI		SN	MONTH	ANES	LOI
38	2	1	0		10	1	3	8
43	2	1	0		57	3	3	8
58	3	1	0		14	1	4	8
59	3	1	0		29	2	4	8
1	1	1	3		44	2	4	8
35	2	1	3		52	3	4	8
40	2	1	3		5	1	1	9
16	1	5	3		37	2	3	9
11	1	1	4		48	3	2	10
34	2	1	4		61	3	3	10
56	3	1	4		21	2	5	10
26	2	3	4		36	2	5	10
54	3	5	4		7	1	2	11
25	2	1	5		47	3	3	11
41	2	1	5		17	1	4	11
4	1	2	5		6	1	1	12
8	1	2	5		27	2	2	12
30	2	2	5		60	3	4	14
28	2	3	5		23	2	1	16
39	2	3	5		12	1	3	16
2	1	1	6		42	2	3	17
20	1	2	6		46	3	4	17
53	3	3	6		31	2	4	18
55	3	4	6		50	3	5	18
9	1	5	6		3	1	2	19
22	2	5	7		24	2	5	19
45	2	5	7		15	1	2	20
51	3	5	7		49	3	5	84
19	1	2	8		13	1	4	85
33	2	2	8		18	1	1	280
					32	2	2	605

Figures 1 and 2 are the histogram and probability plot for all 61 LOI times during the baseline three-month period. These 61 cases include the four complicated cases. Both the histogram and probability plot show the distribution to be severely skewed to the right as might have been anticipated. Note the clear demarcation on the probability plot between the 57 uncomplicated cases and the four complicated ones.

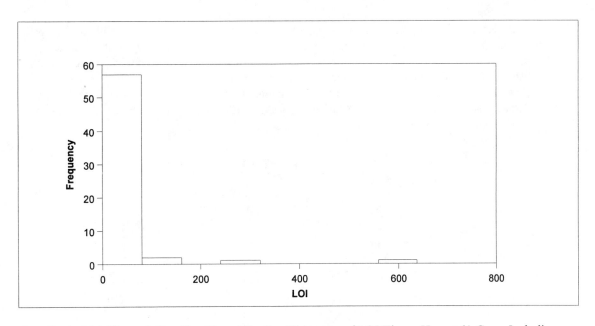

Case Study 10.1 Figure 1   Baseline Three Months: Histogram of LOI Times, Hours; 61 Cases Including Four with Complications (Data in File ZLT3)

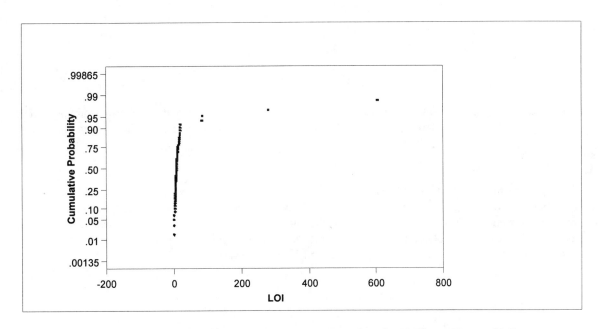

Case Study 10.1 Figure 2   Baseline Three Months: Probability Plot of LOI Times, Hours; 61 Cases Including Four with Complications (Data in File ZLT3)

Figure 3 is the Xbar and s chart on LOIs for the baseline three months, subgrouped by ANES and using 2.5-sigma limits (since there are only five anesthesiologists). The chart is shown only for its value in demonstrating that the Xbar chart and the s chart go up and down together (are "in phase"), which occurs when a distribution is badly skewed to the right. This shows the Xbar and s chart to be invalid because there is too great a departure from a normal distribution. For the rare case where the distribution is badly skewed to the left, the s chart goes down when the Xbar chart goes up ("180 degrees out of phase"). Note that the LCL on Xbar has been set to zero.

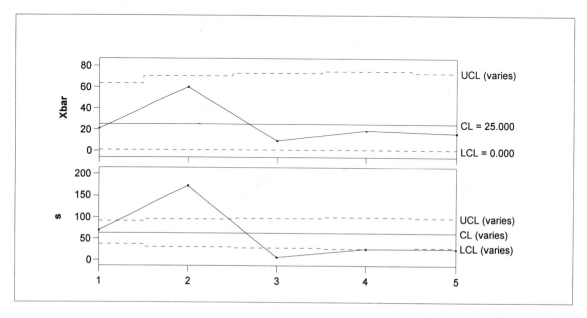

Case Study 10.1 Figure 3   Baseline Three Months: Xbar and s chart of LOI Times Subgrouped by ANES #, 2.5-Sigma Limits; 61 Cases Including Four with Complications (Data in File ZLT3)

After deleting the four complicated cases, the 57 uncomplicated cases for the baseline three-month period have been saved as ZUT3. The histogram and probability plot, Figures 4 and 5, both are typical for a bimodal distribution. This is because the distribution includes anesthesiologists operating with different LOI averages. However, the probability plot is now sufficiently close to being a straight line so that the distribution may be considered near-normal and used for control chart analysis.

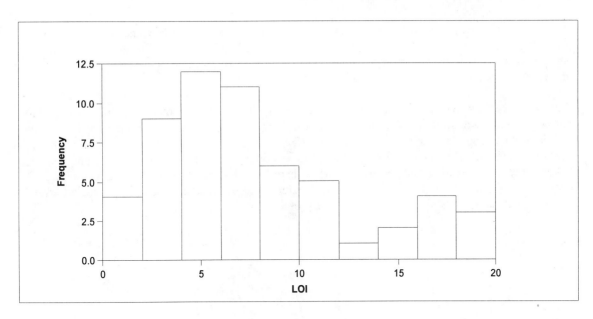

Case Study 10.1 Figure 4   Baseline Three Months: Histogram of LOI Times, Hours; 57 Uncomplicated Cases (Data in File ZUT3)

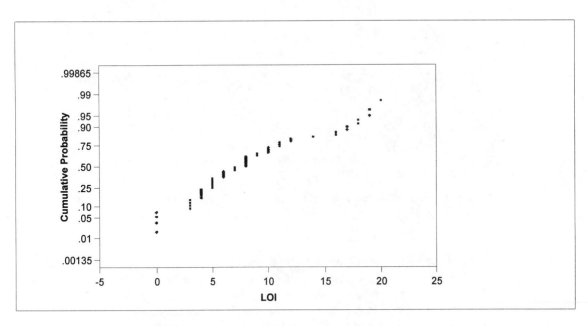

Case Study 10.1 Figure 5   Baseline Three Months: Probability Plot of LOI Times, Hours; 57 Uncomplicated Cases (Data in File ZUT3)

Before using an Xbar and s chart to compare the five anesthesiologists for the baseline three months, it is good practice to study each of the five in time order with an I chart. The objective here is to see if there is any glaring peculiarity that would render the Xbar and s chart invalid, such as one anesthesiologist trending up over the three months while the others trended down.

There are not enough data for a good analysis of each anesthesiologist over time; the number of procedures ranged between 9 and 16 for the five anesthesiologists. No serious special-cause variation was found. The only indication of special-cause variation was for ANES #1, Figure 6, with one point barely above the upper control limit (UCL). This can be ignored since it was only out by 0.2 hours and the times are only recorded to the nearest hour; recall that all you are looking for here are gross indications that would render the comparison of anesthesiologists invalid. However, Figure 6 suggests that the second half of the observations may be significantly lower than the first half. (See problem 2.)

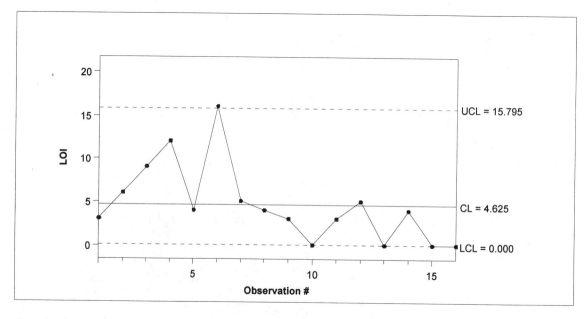

Case Study 10.1 Figure 6 Baseline Three Months: I Chart of LOI Times, Hours; ANES #1 (Data in File ZUT3)

Figure 7 is the Xbar and s chart with 2.5-sigma limits for the baseline 57 uncomplicated LOI times subgrouped by ANES #. Now that the complicated procedures have been deleted, the Xbar chart and the s chart are no longer in phase; the distribution is near-normal so the charts can be expected to be valid. Note that the s chart has no points outside the 2.5-sigma limits, so the Xbar chart may be interpreted with confidence. The Xbar chart shows the average LOI for ANES #1 to be so low that it could not reasonably be attributed to random chance. The subgroup sizes, shown on the Xbar chart, are smaller than would be desirable for the formal evaluation of whether or not special-cause variation exists between the five anesthesiologists, but they are perfectly adequate for the present purpose—to look for guidance on how to improve process performance.

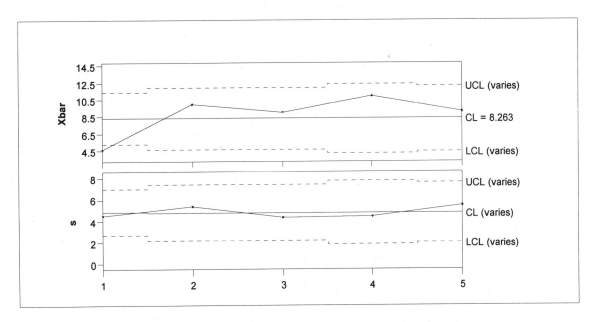

Case Study 10.1 Figure 7 Baseline Three Months: Xbar and s chart of LOI Times, Hours, Subgrouped by ANES, 2.5-sigma Limits; 57 Uncomplicated Cases (Data in File ZUT3)

Table 1 disclosed that ANES #1 had all four of the cases where the extubation was done in the OR. The p chart in Figure 8 profiles the anesthesiologists on the percentage of LOI times that were zero for the baseline three months using 2.5-sigma limits. ANES #1 is above the UCL. The subgroup size for ANES #1 is 16 procedures. If this outage were to be used for a formal evaluation of whether or not special-cause variation existed between the anesthesiologists, the subgroup size for ANES #1 would have to be no less than $4/pBar = 4/0.070 = 57.1$, so no interpretation would be possible. (The advanced reader could use binomial probability limits.) However, for the purpose of guidance on process improvement, this result confirms the Xbar and s chart finding: process improvement efforts should concentrate on the differences in LOI times between ANES #1 and the others.

At their second April meeting, the anesthesiologists reviewed the first-quarter data and decided to use the results of ANES #1 as a benchmark. The initial three months would be used as a baseline to see whether improvement could be made. They agreed that this would involve extubating in the operating room where this was appropriate and using other guidelines to extubate as early as was reasonable for all but the complicated cases.

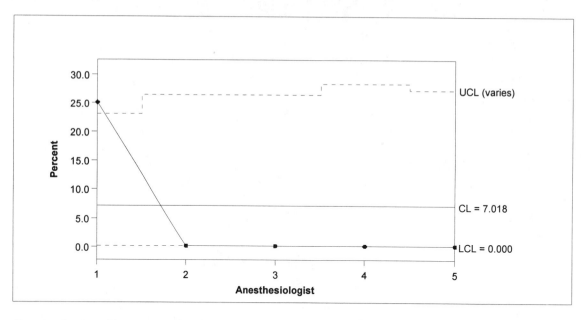

Case Study 10.1 Figure 8   Baseline Three Months: p Chart on Percent of Extubations Performed in the OR, Subgrouped by ANES, 2.5-Sigma Limits (Data in File ZUT3)

Comparison of Baseline and Final: Data

(Data are in file ZUAT2.) The second part of this study compares months 1 through 3 with months 5 through 7 for ANES #2 through #5 to evaluate whether those anesthesiologists made an improvement between the two eras. The file ZUAT2 has data similar to that in Table 1 for all uncomplicated cases, months 1 through 7. To be able to see whether improvement was made by ANES #2 through ANES #5, pooled together, ZUAT also identifies two eras: the baseline three months and the final three months excluding ANES #1. Data are also included in that file to make a p chart on the percent of extubations in the OR for each of the two eras for ANES #2 through ANES #5 pooled together.

Comparison of Baseline and Final: Analysis, Results, and Interpretation

ANES #2 through ANES #5 are pooled together. Ignoring month 4, the transition month, their LOIs for the baseline era are compared to their LOIs for the era of the final three months. First, a p chart is used to compare the percentage of extubations performed in the OR in the two time eras. Finally, an Xbar and s chart on LOI times is used to compare the two eras. In each case 1.5-sigma limits are used on a control chart where the objective is a formal evaluation to see whether there is special-cause variation between the two eras.

The p chart, Figure 9, shows outages that would indicate improvement had been made. However, the minimum subgroup size required to believe these results for process evaluation is not met. The minimum size needed is 4/pBar = 4/0.067 = 60 procedures. The actual subgroup sizes are only 41 and 49. The improvement from zero extubations in the OR during the first era to six in the second era may have been due to special-cause variation, but there is so little data that we must accept that it could well have been simply random chance—just common-cause variation.

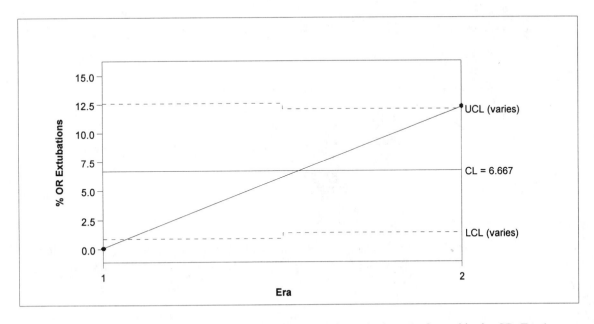

Case Study 10.1 Figure 9 p Chart, Percent of Uncomplicated Extubations Performed in the OR, Era 1 versus Era 2, ANES #2 through ANES #5, Pooled Together, 1.5-Sigma Limits

The Xbar and s chart demonstrates more power than the p chart in this case. In Figure 10, the s chart is in control so the Xbar chart has valid limits. The subgroup sizes are large enough and the improvement in average LOI shown on the Xbar chart constitutes evidence that there was special-cause variation between the two eras; there was improvement by anesthesiologists 2 through 5 after benchmarking anesthesiologist 1.

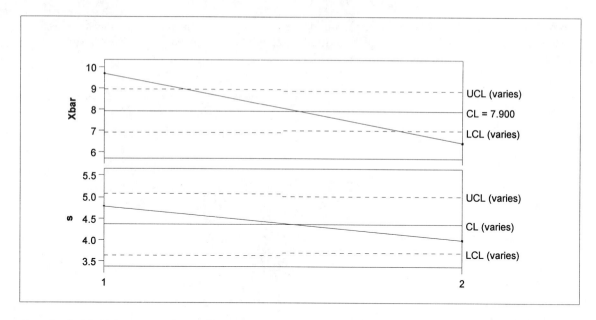

Case Study 10.1 Figure 10 Xbar and s chart of Uncomplicated LOIs Subgrouped by Era (Baseline Three
Months versus Final Three Months), ANES #2 through ANES #5, Pooled
Together, 1.5-Sigma Limits

Lessons Learned

- The histogram and probability plot will show when the distribution is severely skewed and may provide
  guidance on how to select a subset of the data that is near-normal.

- The Xbar and s chart will also indicate when the data are badly skewed to the right by going up and
  down together.

- When applicable, it is good practice to make an I chart on each subgroup over time in addition to
  making an Xbar and s chart.

- Sometimes variables data can advantageously be converted to attribute data to obtain another
  perspective on a problem.

- When the objective is formal evaluation for evidence of special-cause variation, the guidelines for the
  quantities of data and the T-sigma limits should be followed.

Management Considerations

The primary reason for the success of this improvement effort is that it was the anesthesiologists' own idea
and they took ownership of it. Under these circumstances, the deck is stacked for victory.

Problems

1. Refer to Figure 6. Using data file ZUT3, verify that there was no special-cause variation over time for the first three months for ANES #2 through ANES #5. Can it be stated that there is only common-cause variation?

2. Is there evidence of special-cause variation between the first and second halves of the data in Figure 6?

Appendix 1 Tests for Evidence of Special-Cause Variation

Tests for evidence of special-cause variation when looking at time-ordered data warrants some discussion. The AT&T (formerly Western Electric) *Statistical Quality Control Handbook* [Western Electric, 1956] and Nelson's rules [1984] are probably the most often quoted on this. Some of these tests have been discussed in brief in Chapter 4.

Western Electric/AT&T Rules

Western Electric/AT&T [1956] suggests considering only one-half of the control band at a time, that is, the area between the centerline and one of the control limits. They then divide the area into three equal zones, A, B, and C, as illustrated in Figure 1. Western Electric/AT&T [1956, p. 25] states that "since the control limits are 3 sigma limits, each of the zones is one sigma in width." Then [1956, pp. 25 – 26]

> The pattern is unnatural if any of the following combinations are formed in the various zones:

Test 1.   A single point falls outside of the 3 sigma limit (beyond Zone A).
Test 2.   Two out of three successive points fall in Zone A or beyond.
Test 3.   Four out of five successive points fall in Zone B or beyond.
Test 4.   Eight successive points fall in Zone C or beyond.

These tests are illustrated in Chapter 4.

0.00135	
———————	Upper Control Limit
0.0214 Zone A	
————	
0.1359 Zone B	
————	
0.3413 Zone C	
———————	Centerline

Appendix 1 Figure 1 Zones and the Probability of a Point Being in the Zone, Normal Distribution Assumed

Even these simple criteria get complicated by Western Electric/AT&T. They state [1956, p. 28]:

> The above tests apply when the two control limits on the chart are at reasonably similar distances above and below the centerline. On an $\overline{X}$ chart the control limits are always symmetrical, but on an R chart or p-chart the control limits are sometimes unsymmetrical. Unsymmetrical limits may require a slight change in the application of the tests.

Limits for s would be in the same category since the distributions of standard deviations are also unsymmetric.

As for p charts and various other control charts, Western Electric/AT&T suggests [1956, p. 183]:

> . . . it is possible to calculate special tests for p-charts or other charts whose control limits may at times be unsymmetrical. First find the probability associated theoretically with each third of the control band, and calculate tests which will result in the desired total probability of getting a reaction to the tests.

Western Electric/AT&T suggests that the following patterns should be watched for in addition to patterns of instability [1956, p. 29]:

> (1) Stratification . . .
> Consider that stratification exists when 15 or more consecutive points fall in Zone C, either above or below the centerline.
>
> (2) Mixture . . .
> Consider that mixture exists when the chart shows 8 consecutive points on both sides of the centerline with none of the points falling in Zone C.
>
> (3) Systematic variation
> The presence of a systematic variable in the process is indicated if a long series of points are high, low, high, low without any interruption in this regular sequence.
>
> (4) Tendency of one chart to follow another
> Two variables are likely to be related to each other if a long series of points on their respective patterns move up and down in unison.
>
> (5) Trends may be indicated by:
> (a) x's on one side of the chart followed by x's on the other.
> (b) a series of consecutive points without a change in direction.

The last three items are not explicitly defined in Western Electric/AT&T.

Nelson's Rules

For means of normally distributed data (Xbar chart) and for normally distributed individual values (I chart), Nelson [1984, p. 238] suggests 8 tests that are sometimes referred to as Nelson's rules:

Test 1.   One point beyond zone A. (Same as Western Electric/AT&T test 1.)

Test 2.   Nine points in a row in zone C or beyond.

Test 3.   Six points in a row (including endpoints) steadily increasing or decreasing.

Test 4.   Fourteen points in a row alternating up and down. (Similar to Western Electric/AT&T Systematic Variation Test.)

Test 5.   Two out of three points in a row in zone A or beyond. (Same as Western Electric/AT&T test 2.)

Test 6.   Four out of five points in a row in zone B or beyond. (Same as Western Electric/AT&T test 3.)

Test 7.   Fifteen points in a row in zone C (above and below centerline). (Same as Western Electric/AT&T Stratification Test.)

Test 8.   Eight points in a row on both sides of the centerline with none in zone C. (Same as Western Electric/AT&T Mixture Test.)

Note that Nelson calls for nine points below the centerline (test 2) and Western Electric/AT&T calls for eight (their test 4 above). These tests are illustrated in Chapter 4.

Nelson [1984, p. 237] also comments:

> Test one, three, and four can be used with p, np, c, and u charts. If the distributions are close enough to being symmetrical, test two can also be used with these charts. Use binomial or Poisson tables to check specific situations.

Most software packages use Nelson's rules or those by Western Electric/AT&T.

Runs

The theory of runs is occasionally applied to control charts. Two types of runs are discussed.

1. *Runs above the centerline or runs below the centerline.* These runs consider the consecutive points above the centerline (or below the centerline), ignoring the points exactly equal to the centerline, as a run. The total number of runs are counted and/or the length of the longest run is counted.

2. *Runs up and runs down.* A succession of increases in value is considered a run up; of decreases is a run down. Some authors, including this text, count both end points to determine the length of the run. However, some authors [Duncan, 1986] include only one end point since they count the number of increases (or

decreases), not the points producing the increases or decreases. The total number of runs up or down are counted and/or the length of the largest run is counted.

Formally, the tests for runs can get rather complicated because they depend on the number of points plotted as well as the α-risk (the probability of the run happening by random chance) chosen.

Runs above the Centerline or Runs below the Centerline

To check the number of runs to see if there are too few runs, the number of points both above the centerline and below the centerline are counted. The smaller of the two counts is called r; the larger is called s. The total number of runs is counted. Using Tables 1 and 2 (provided by Duncan, 1986, p.1027), the limiting value for the total number of runs is found. If the counted number of runs is less than that shown in Table 1, the probability is less than 0.005 that the number of runs could have occurred by pure chance. Note that this table was derived for counts above or below the median. However, for a normal distribution, the median is equal to the mean. If desired, Table 2 [Duncan, 1986, p. 1027] could be used for the limiting value of runs with a probability of less than 0.05. This will, however, bring the probability of reaction to tests higher than was suggested by AT&T.

Appendix 1 Table 1 Testing Randomness of Groupings in a Sequence of Alternatives (Probability of an Equal or Smaller Number of Runs Than That Listed is p = 0.005)

s = cases on one side of the centerline (the side with the larger number of runs)
r = cases on the other side of the centerline, r ≤ s

s\r	6	7	8	9	10	11	12	13	14	15	16	17	18	19	20
6	2														
7	2	3													
8	3	3	3												
9	3	3	3	4											
10	3	3	4	4	5										
11	3	4	4	5	5	5									
12	3	4	4	5	5	6	6								
13	3	4	5	5	5	6	6	7							
14	4	4	5	5	6	6	7	7	7						
15	4	4	5	6	6	7	7	7	8	8					
16	4	5	5	6	6	7	7	8	8	9	9				
17	4	5	5	6	7	7	8	8	8	9	9	10			
18	4	5	6	6	7	7	8	8	9	9	10	10	11		
19	4	5	6	6	7	8	8	9	9	10	10	10	11	11	
20	4	5	6	7	7	8	8	9	9	10	10	11	11	12	12

Source: From *Quality Control and Industrial Statistics*, Fifth Edition, by A. Duncan, p. 1027. Copyright © 1986 Richard D. Irwin, Inc. Reprinted with permission.

Appendix 1 Table 2 Testing Randomness of Groupings in a Sequence of Alternatives (Probability of an Equal or Smaller Number of Runs than That Listed is p = 0.05)

s = cases on one side of the centerline (the side with the larger number of runs)
r = cases on the other side of the centerline, r ≤ s

s\r	6	7	8	9	10	11	12	13	14	15	16	17	18	19	20
6	3														
7	4	4													
8	4	4	5												
9	4	5	5	6											
10	5	5	6	6	6										
11	5	5	6	6	7	7									
12	5	6	6	7	7	8	8								
13	5	6	6	7	8	8	9	9							
14	5	6	7	7	8	8	9	9	10						
15	6	6	7	8	8	9	9	10	10	11					
16	6	6	7	8	8	9	10	10	11	11	11				
17	6	7	7	8	9	9	10	10	11	11	12	12			
18	6	7	8	8	9	10	10	11	11	12	12	13	13		
19	6	7	8	8	9	10	10	11	12	12	13	13	14	14	
20	6	7	8	9	9	10	11	11	12	12	13	13	14	14	15

Source: From *Quality Control and Industrial Statistics*, Fifth Edition, by A. Duncan, p. 1027. Copyright © 1986 Richard D. Irwin, Inc. Reprinted with permission.

To check the length of the longest run above or below the centerline, Duncan [1986, p. 1029] gives a table, repeated here in Table 3. If there are 20 total points, a run of 8 points gives an α-risk of 0.01 or a run of 9 points gives an α-risk of 0.001, enough to indicate special-cause variation. This is consistent with the rule of thumb used by many quality professionals that a run of 8 (Western Electric/AT&T) or a run of 9 (Nelson's rules) above (or below) the centerline is enough to give an indication of special-cause variation.

Appendix 1 Table 3 Limiting Values for Lengths of Runs on Either Side of the Centerline of *n* Points
(Probability of Getting at Least One Run of Specified Size or More)

n	0.05	0.01	0.001
10	5	-	-
20	7	8	9
30	8	9	-
40	9	10	12
50	10	11	-

Source: From *Quality Control and Industrial Statistics*, Fifth Edition, by A. Duncan, p. 1029. Copyright ©
1986 Richard D. Irwin, Inc. Reprinted with permission.

Runs Up and Runs Down

The same analysis used for the number of runs above or below the centerline can be applied to the number
of runs up and runs down. The number of individual increases and the number of individual decreases are
counted. Let r be the smaller count, s the larger. Count the total number of runs up or down. This count is
compared to the number in Table 1 (or Table 2) as before.

To check the length of the largest run up or run down, Table 4 is derived from Duncan [1986, p. 1029]
using his definition of a run up as the number of increases. If it is desirable to use a maximum α-risk of
around 0.05, the right side column of Table 4 would be used. For 20 points, a run of 5 or more increases (6
or more points including endpoints) would have a probability of 0.0355. This suggests special-cause
variation. This is consistent with the rule of thumb used by many quality professionals that a run of 6 or
more points (including endpoints) indicates special-cause variation.

Appendix 1 Table 4 Limiting Lengths of Runs Up and Down in a Series of *n* Numbers

n	Probability Equal to or Less Than 0.0032		Probability Equal to or Less Than 0.0567	
	Run	Probability of an Equal or Greater Run	Run	Probability of an Equal or Greater Run
4	4	0.0028	4	0.0028
5	5	0.0004	4	0.0165
6	5	0.0028	4	0.0301
7	6	0.0004	4	0.0435
8	6	0.0007	4	0.0567
9	6	0.0011	5	0.0099
10	6	0.0014	5	0.0122
11	6	0.0018	5	0.0146
12	6	0.0021	5	0.0169
13	6	0.0025	5	0.0193
14	6	0.0028	5	0.0216
15	6	0.0032	5	0.0239
20	7	0.0006	5	0.0355
40	7	0.0015	6	0.0118
60	7	0.0023	6	0.0186
80	7	0.0032	6	0.0254
100	8	0.0005	6	0.0322
200	8	0.0010	7	0.0085
500	8	0.0024	7	0.0215
1,000	9	0.0005	7	0.0428
5,000	9	0.0025	8	0.0245

Source: From *Quality Control and Industrial Statistics*, Fifth Edition, by A. Duncan, p. 1029. Copyright © 1986 Richard D. Irwin, Inc. Reprinted with permission.

<u>Caution</u>

Additional tests increase the α-risk. Montgomery [1997, p. 150] warns that using additional criteria should be done with caution "as an excessive number of false alarms can be harmful to an effective SPC program. Finally, as more supplemental rules are applied to the chart, the decision process becomes more complicated, and the inherent simplicity of the Shewhart Control Chart is lost."

Appendix 2 The Perils of Pooling

Pooling of separate streams of data increases the amount of data available for a single analysis and hence increases the power of that analysis. However, before two separate streams of data may be pooled, it must be shown that there is no special-cause variation between the two streams, that is, that they are "like" streams of data.

Pooling of "unlike" streams of data can be disastrous, as will be shown in the examples below. It is sometimes obvious that pooling would be foolish. Consider the case of the man with one foot on a hotplate and the other foot on a block of ice. "On the average," or perhaps we should say "on the *apparent average*," he may be quite comfortable. The apparent average may be calculated, but it has no meaning. As another case, consider reporting the apparent average height of the people in a community, ignoring age and gender; it is unlikely that you would make such a mistake.

Improper pooling is sometimes the result of knowingly combining streams of data that should not be pooled. More often, improper pooling is the result of not even considering that separate streams might exist. The improper pooling of unlike streams is usually done in all innocence, without any thought that there may be separate streams within the data set at hand. This "hidden" pooling is illustrated in the following two examples.

EXAMPLE 1 Surgery Complication Rates

The Press Release

Thoracic deep-wound procedures play an important role at East Hospital. For the past two years the hospital has carefully monitored all aspects of these procedures, including complications. At the end of year 2 a press release was issued stating that these thoracic procedures had doubled in year 2 and that the complication rate had dropped by 1%.

The Data

(Data are in file ZTHOR.) Table 1 shows the complication data for thoracic deep-wound surgery procedures for years 1 and 2. Note that what others might just call the "complication rate" is called the *apparent complication rate* in this table to emphasize that, although it may be calculated, it has no meaning since it comes from pooling unlike streams of data—male and female.

Appendix 2 Table 1 Surgery Complication Data, Years 1 and 2

	Year 1		
	Male	Female	Total
Uncomplicated Cases	91	48	139
Complicated Cases	10	30	40
	10/101 = 9.9% Male Complication Rate	30/78 = 38.5% Female Complication Rate	40/179 = 22.3% Apparent Complication Rate
Total	101	78	179

	Year 2		
	Male	Female	Total
Uncomplicated Cases	249	30	279
Complicated Cases	31	46	77
	31/280 = 11.1% Male Complication Rate	46/76 = 60.5% Female Complication Rate	77/356 = 21.6% Apparent Complication Rate
Total	280	76	356

Analysis, Results, and Interpretation

Table 1 shows that from year 1 to year 2 the apparent complication rate dropped from 22.3% to 21.6%. However, both the male and female complication rates increased! The male rate increased from 9.9% to 11.1% while the female complication rate suffered a very large increase—from 38.5% to 60.5%.

The literature in recent years has contained numerous references to the fact that females have higher CABG surgery mortality rates than males. It would have been reasonable to suspect that there might be a gender effect in the complication rate for thoracic deep-wound surgical procedures. A reasonable first step in the data analysis for this example is to see whether the proportion of female patients to males was constant over the two-year period. Figure 1 is a p chart on the percentage of female patients subgrouped by year. It may be seen that the decrease in females (and the accompanying increase in males) was of such a large magnitude that it could not reasonably be attributed to random chance. The gender mix is a special cause of variation between years 1 and 2.

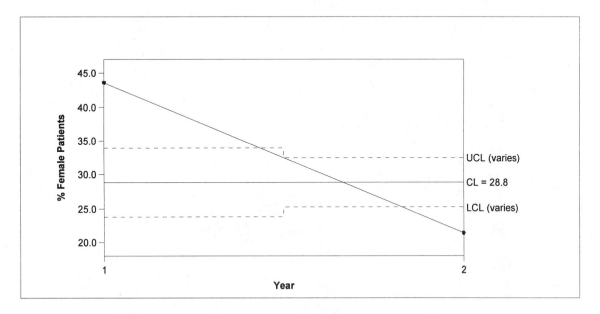

Appendix 2 Figure 1 p Chart on Percentage of Female Patients, Subgrouped by Year, 1.5-Sigma Limits

Particularly in view of the changing gender mix for the two years, it is essential that the two genders have like complication rates for the two years. This is not the case. Figure 2 shows that the female complication rate in year 1 is about four times as high as for males; for year 2, Figure 3 shows the female rate to be about six times as high as for males. In each year the high female complication rate is a special cause of variation. The two streams of data (for the two genders) must not be pooled. The *apparent complication rate* is shown to have no meaning, so attempting to analyze its change from year 1 to year 2 would be foolish.

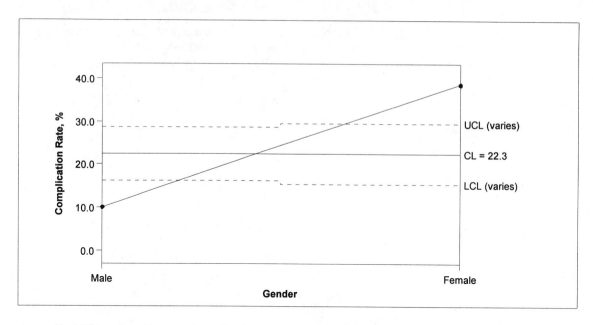

Appendix 2 Figure 2 p Chart on Complication Percentage for Year 1, Male versus Female, 1.5-Sigma Limits

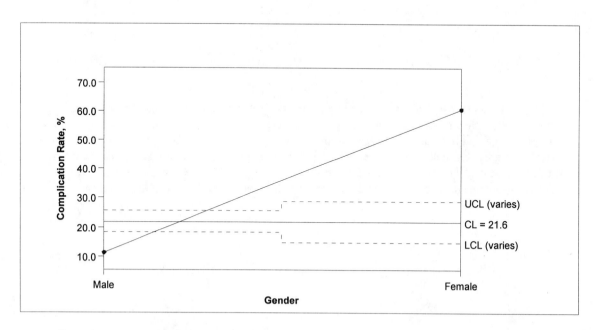

Appendix 2 Figure 3 p Chart on Complication Percentage for Year 2, Male versus Female, 1.5-Sigma Limits

From Figure 4 it is seen that the increase in the male complication rate in year 2 could not be shown to be special-cause variation and so is attributed to random chance. Figure 5 shows that the large increase in the female complication rate in year 2 was special-cause variation. The process deteriorated badly in the second year and was not just random chance.

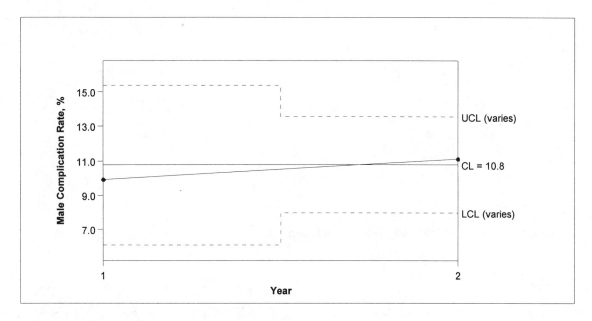

Appendix 2 Figure 4 p Chart on Male Complication Rate, Year 1 versus Year 2, 1.5-Sigma Limits

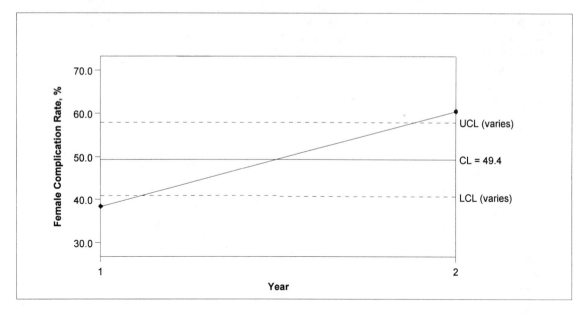

Appendix 2 Figure 5 p Chart on Female Complication Rate, Year 1 versus Year 2, 1.5-Sigma Limits

Lessons Learned

You must always hunt for possibilities of hidden pooling. Experts in the process should be asked for all possible methods of subgrouping the data that might be relevant. Further, a systematic approach to the analysis of data is required, as shown in this example.

Indirect Standardization, Another Analysis Method

(*Note*: This section is supplementary advanced material that may be skipped by the reader without any negative impact on understanding or using the rest of the book.)

In the fields of biostatistics and epidemiology, problems are sometimes encountered where a comparison being sought is clouded or "confounded" by another factor. Methods of standardization may then be used to try to adjust for that confounding factor. The reader is referred to Pagano [2000] for a review of standardization methods. This application is restricted to the use of what is called *indirect standardization*.

Some new terminology must be introduced for the use of standardization. Let the word "subgroups" be restricted here to the two or more subpopulations to be compared. For another factor to obscure the analysis, that confounding factor must exist at two or more "levels" or "strata." In this problem of surgery complication proportion, the two subgroups to be compared are "year 1" and "year 2." The confounding factor is "gender," which has two strata, "male" and "female."

Two types of proportions were considered above:

1. The apparent complication proportion for each year (0.223 and 0.216), sometimes called the crude proportion.
2. The four stratum-specific proportions (e.g., 0.099 for male, year 1).

If there had been no confounding factor (i.e., if the two genders had suffered the same complication rate) pooling would have been appropriate and a control chart on the crude complication proportions would have told the whole story: Was any change in proportion complicated between year 1 and year 2 a special cause or should it be attributed to natural variation? However, since gender was a confounding factor, the crude complication proportions do not provide a fair comparison between the two years. Such a comparison in the face of a confounding factor is the province of standardization.

The stratum-specific proportions are always valid and may be extremely useful to guide improvement efforts. It can be seen in Figures 4 and 5 that the small increase in male complication proportion between the two years may reasonably be attributed to random chance, but that the large increase in female complication proportion is a special cause. Here is a serious problem and the opportunity to improve!

Before proceeding with the indirect standardization method, it should be emphasized that the sole purpose of standardization is to provide a valid comparison of two or more subgroups that contain stratified data confounding the crude proportions. Such confounding could lead to an invalid comparison. The conditions under which standardization is useful and the limitations of the results will be discussed after the development of the method.

First, recall that the total *observed* complication proportion subgrouped by year has been completely discredited as a source of information in the preceding section. Nevertheless, this statistic will be examined here with a p chart as a point of departure. From Table 1 note that the total number of observed complications is 40 out of 179 procedures in year 1 and 77 out of 356 in year 2. Figure 6 is the control chart for observed complication proportion subgrouped by year, using 1.5-sigma limits. Here it may be seen that the small decrease in observed complication proportion from year 1 to year 2 may be attributed to random chance.

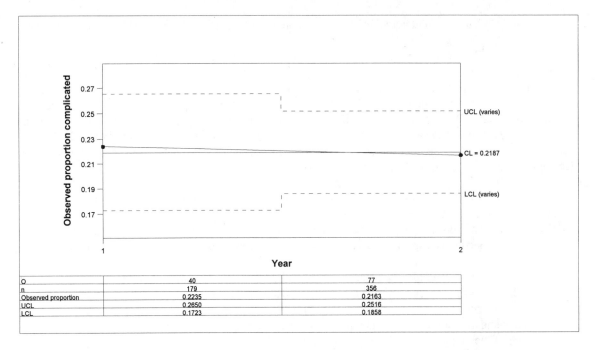

		40		77
O		40		77
n		179		356
Observed proportion		0.2235		0.2163
UCL		0.2650		0.2516
LCL		0.1723		0.1858

Appendix 2 Figure 6 p Chart for Observed Proportion Subgrouped by Year, 1.5-Sigma Limits

Next, a p chart will be developed to study the total *expected* complication proportion, subgrouped by year. Standardization employs an arbitrarily defined standard population, which has been chosen here as the sum of the two yearly subgroups. For this indirect standardization method, the expected complication proportion for each stratum (each gender in this case) is the standard population complication proportion for that stratum. Using the Table 1 data, the expected complication proportion for the standard population is

$$(10 + 31)/(101 + 280) = 0.10761$$

for males and

$$(30 + 46)/(78 + 76) = 0.49351$$

for females. The expected count must now be found for each stratum for each year. This count is the product of the applicable gender complication proportion above and the size of the stratum for that gender and year. For year 1, the expected count is

$$(0.10761)(101) = 10.869$$

for males and

$$(0.49351)(78) = 38.494$$

for females, giving a total expected complication count for year 1 of 49.363. Similarly, it may be verified that the expected complication count for year 2 is 67.638. Figure 7 is the control chart for expected proportion subgrouped by year, using 1.5-sigma limits. Here it may be seen that the decrease in expected complication proportion from year 1 to year 2 is special-cause variation, too great to be attributed to random chance. Comparing this chart with Figure 6, it can be seen that the decrease in observed proportion was far less than should have been expected. It will next be shown that this failure to meet expectation is disclosed by the indirect standardization method.

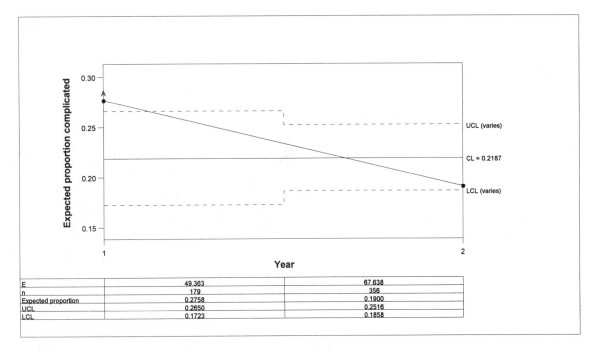

Appendix 2 Figure 7 p Chart for Expected Proportion Subgrouped by Year, 1.5-Sigma Limits

Using the indirect standardization method, each observed total yearly complication count is multiplied by a factor to adjust it for the confounding effect of gender. The adjustment factor for each year is the ratio of the overall expected complication proportion to the expected yearly complication proportion. The numerator for both years is the p chart centerline in Figure 7, 0.2187. The denominator for the first year is the expected count for year 1 divided by the total number of procedures for year 1, that is, 49.363/179 = 0.2758, which may be verified in the tabular portion of Figure 7. For year 1 the adjustment factor is 0.2187/0.2758 = 0.7930, so the adjusted count is 40(0.7930) = 31.72. Similarly, it may be seen that the adjustment factor for year 2 is 1.1511 and the adjusted count is 88.63.

Figure 8 is the p chart for *adjusted* proportion subgrouped by year, using 1.5-sigma limits. Here it may be seen that the increase in adjusted complication proportion from year 1 to year 2 is special-cause variation, too great to be attributed to random chance. The indirect standardization method yielded the expected result.

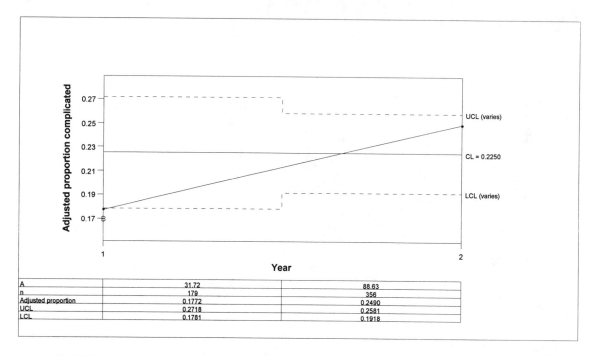

A	31.72	88.63
n	179	356
Adjusted proportion	0.1772	0.2490
UCL	0.2718	0.2581
LCL	0.1781	0.1918

Appendix 2 Figure 8 p Chart for Adjusted Proportion Subgrouped by Year, 1.5-Sigma Limits

The use of standardization requires that there be no significant interaction between the subgroups and the strata (between year and gender here). This freedom from interaction is a subjective evaluation made by determining that the subgroups have plots of observed proportion as a function of stratum that are more or less similar in shape. Figure 9 shows this requirement to be satisfied so that the analysis above is validated.

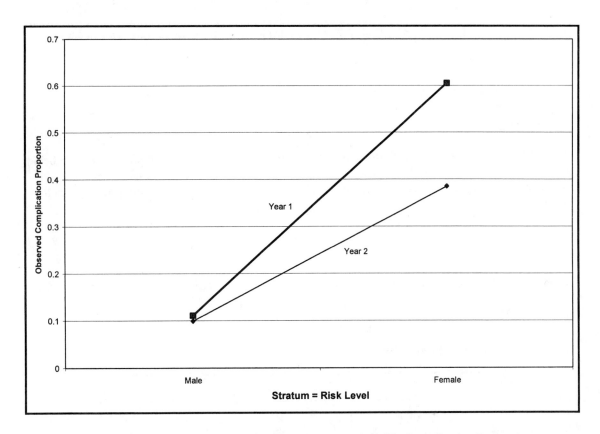

Appendix 2 Figure 9   Plots for Each Year of Observed Stratum-Specific Complication Proportion as a
Function of Stratum

Before extolling the virtues of standardization too highly, the severe limitations of the method must be
considered.

1.   There is no real process that corresponds to the adjusted counts or adjusted proportions.
2.   There is no applicability of the standardized results to anything beyond the comparison of
these particular subgroups with this particular data. Absolutely no generalization may be
made.
3.   The standardization method provides no information to improve the process. This must be
done by use of the stratum-specific control charts, Figures 4 and 5. This may be viewed as
an application of Deming's observation that to improve the river, you must work on the
streams feeding it. Here the two separate streams are the two genders.

These limitations notwithstanding, if the sole objective is a comparison of two or more subgroups taking
into account a confounding factor, indirect standardization will accomplish that goal.

This control chart approach takes the typical standardization study one step further by using control charts
on the results to determine whether the comparisons indicate special-cause variation or only differences that

321

can reasonably be attributed to random chance. This same method may be used for the next example on cesarean sections. □

EXAMPLE 2 Caesarian Section Rates

The Press Release

The obstetrics department at West Hospital has been only too conscious of the widespread attention being given to cesarean section rates throughout the nation. At the end of two years of closely monitoring hospital performance, a press release was issued stating that the C-section rate had dropped by 6% in the most recent year.

The Data

Table 2 shows the C-section data for years 1 and 2. Definitions of the various rates are given in Case Study 8.1.

Appendix 2 Table 2 C-Section Data, Years 1 and 2

| | Year 1 | | |
	No Previous C-Section	After C-Section	Total
Vaginal Births This Year	489	32 VBACs	521
C-Sections This Year	70 Primary Cs 70/559 = 12.5% Primary C Rate	56 Repeat Cs 56/88 = 63.6% Repeat C Rate	126 Total Cs 126/647 = 19.5% Apparent C Rate
Total	559 Candidates for Primary Cs	88 Candidates for Repeat Cs	647 Total Live Births

	Year 2		
	No Previous C-Section	After C-Section	Total
Vaginal Births This Year	752	11 VBACs	763
C-Sections This Year	113 Primary Cs 113/865 = 13.1% Primary C Rate	64 Repeat Cs 64/75 = 85.3% Repeat C Rate	177 Total Cs 177/940 = 18.8% Apparent C Rate
Total	865 Candidates for Primary Cs	75 Candidates for Repeat Cs	940 Total Live Births

Analysis, Results, and Interpretation

Table 2 shows that from year 1 to year 2 the apparent C-section rate dropped from 19.5% to 18.8%. However, both the primary rate and the repeat rate increased! The primary C-section rate increased from 12.5% to 13.1% while the repeat C-section rate suffered a very large increase—from 63.6% to 85.3%.

In searching for an explanation to this anomaly, a reasonable place to start is to ask whether the percentage of live births that were candidates for repeat C-sections was stable across the two years. Figure 10 is a p chart on the number of candidates for repeat C-sections (i.e., those with previous C-sections) expressed as a percentage of the number of live births, subgrouped by year. It may be seen that the percentage of candidates for repeat C-sections went down so much (and hence the percentage of candidates for primary C-sections went up so much) that year was a source of special-cause variation. This aspect of the data was not stable over the two-year period.

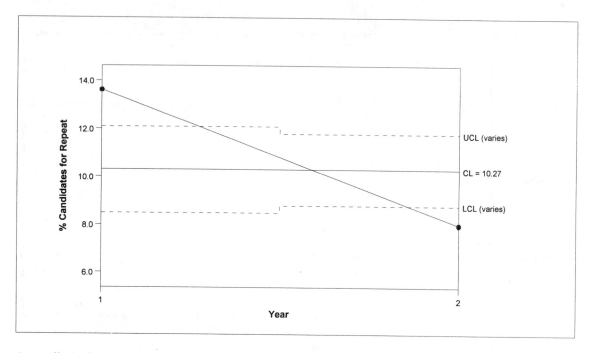

Appendix 2 Figure 10 p Chart on Candidates for Repeat C-Sections as a Percentage of Live Births, Year 1 versus Year 2, 1.5-Sigma Limits

It is common knowledge that the repeat C-section rate runs *much* higher than the primary rate, but the fact that this prohibits the pooling of the two rates apparently has been lost upon those making the press releases. The fact that the "type" of C-section rate (primary vs. repeat) is a source of special-cause variation is verified in Figure 11 for year 1 and in Figure 12 for year 2. In Figure 13 it is seen that the increase in the primary C-section rate in year 2 cannot be shown to be special-cause variation and so is attributed to random chance. Figure 14 shows that the large increase in the repeat C-section rate in year 2 was special-cause variation. The increase in the second year was not just random chance.

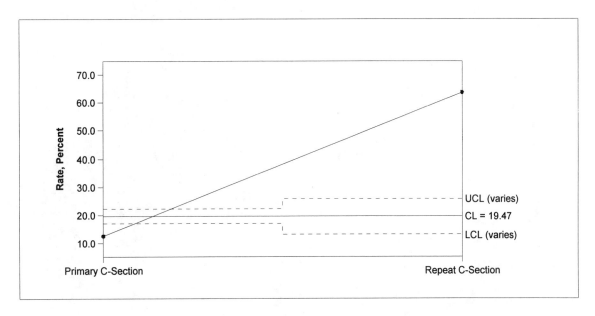

Appendix 2 Figure 11 p Chart on Primary Rate versus Repeat Rate, Year 1, 1.5-Sigma Limits

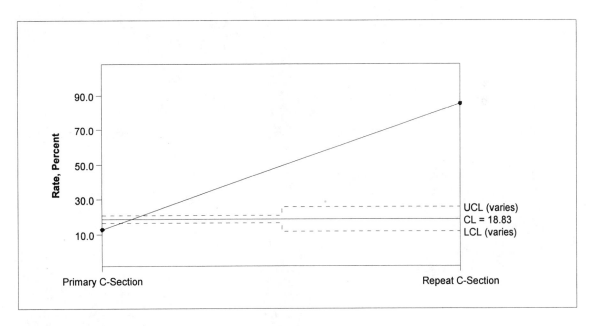

Appendix 2 Figure 12 p Chart on Primary Rate versus Repeat Rate, Year 2, 1.5-Sigma Limits

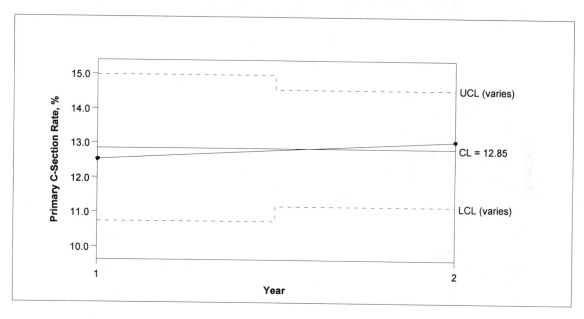

Appendix 2 Figure 13 p Chart on Primary C-Section Rate Subgrouped by Year, 1.5-Sigma Limits

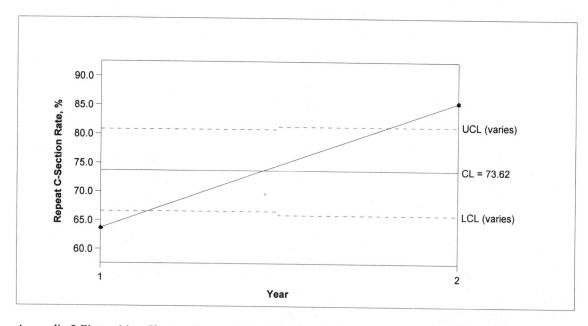

Appendix 2 Figure 14 p Chart on Repeat C-Section Rate Subgrouped by Year, 1.5-Sigma Limits

Lessons Learned

The problem in this example is somewhat more insidious than in the preceding one. Although almost anyone might suspect that males are different than females and therefore that care must be taken in pooling their results, the fact that the same principle holds with common and repeat C-sections has apparently eluded many people.

Management Considerations

It may not be possible for those in managerial positions to be bona fide experts in the field of statistical process control. However, it is essential that they understand the philosophy well enough to *manage* the SPC activity. This implies, of course, that managers must be able to make decisions based on the statistical studies of their experts. But more than this, the managers must be well enough informed to evoke good studies from those statistical experts.

The two examples shown above illustrate how easy it is to unwittingly lie with statistics. Such pitfalls can be largely avoided by having the manager ask a few critical questions:

1. Do those of you who do the statistical analyses and the experts in the subject matter agree that the results adequately reflect *all* conditions under which data might be gathered?
2. Do you all agree that there are no parts of the data (subsets of the data) that, if considered alone, might vary systematically from one another or from the data set as a whole? (male vs. female, young vs. old, different times of the day or days of the week, different surgeons or anesthesiologists, etc.)
3. Do you all agree that there are no "definitional" aspects of the data analysis that might be hiding an invalid pooling of separate streams of data, such as pooling primary C-sections and repeat C-sections? □

Appendix 3 Mathematical Relations of the Variables Control Limit Formulas

The Factor c_4

Many of the mathematical relationships depend on the factor c_4. If sampling from a population having a normal distribution, the standard deviation of the population (σ) may be estimated by

$$\hat{\sigma} = \frac{\bar{s}}{c_4} \tag{1}$$

where

$$c_4 = \sqrt{\frac{2}{n-1}} \, \frac{\left(\frac{n-2}{2}\right)!}{\left(\frac{n-3}{2}\right)!} \tag{2}$$

and n is the subgroup size [ASTM, 1990, p .91]. Values of c_4 may be found in Table 1.

Appendix 3 Table 1 Values of c_4

n	2	3	4	5	6	7	8	9	10	11	12	∞
c_4	.80	.89	.92	.94	.95	.96	.97	.97	.97	.98	.98	$\dfrac{4n-4}{4n-3}$

Control Limits for Xbar for Equal-Size Subgroups

If the sample taken is less than 5% of the population size, the sampling distribution of the mean states that the standard deviation of the averages may be estimated by

$$\hat{\sigma}_{\bar{X}} = \frac{\hat{\sigma}}{\sqrt{n}} \tag{3}$$

The upper control limit (UCL) is taken to be three times the within-subgroup estimate of the standard deviation of Xbar above the overall average; then

$$UCL(\bar{X}) = \bar{\bar{X}} + 3\hat{\sigma}_{\bar{X}}$$

$$= \bar{\bar{X}} + 3\frac{\hat{\sigma}}{\sqrt{n}} \qquad \text{using (3)}$$

329

$$= \overline{\overline{X}} + 3\frac{\overline{s}}{c_4\sqrt{n}} \qquad \text{using (1)}$$

$$= \overline{\overline{X}} + A_3\overline{s}$$

where

$$A_3 = \frac{3}{c_4\sqrt{n}}$$

Similarly, the lower control limit (LCL) is

$$LCL(\overline{X}) = \overline{\overline{X}} - A_3\overline{s}$$

Control Limits for s for Equal-Size Subgroups

When sampling from a population having a normal distribution, the standard deviation of the standard deviation (σ_s) is estimated by

$$\hat{\sigma}_s = \hat{\sigma}\sqrt{1 - c_4^2} \qquad (4)$$

$$= \frac{\overline{s}}{c_4}\sqrt{1 - c_4^2} \qquad \text{using (1)}$$

[ASTM, 1990, p. 93]. Using Shewhart's approach of 3-sigma limits even if the distribution is not normally distributed,

$$UCL(s) = \overline{s} + 3\hat{\sigma}_s = \overline{s} + 3\frac{\overline{s}}{c_4}\sqrt{1 - c_4^2} \qquad \text{using (4)}$$

$$= B_4\overline{s}$$

where

$$B_4 = 1 + \frac{3}{c_4}\sqrt{1 - c_4^2}$$

Similarly,

$$LCL(s) = B_3\overline{s}$$

where

$$B_3 = 1 - \frac{3}{c_4}\sqrt{1 - c_4^2}$$

<u>Control Limits for X</u>

If sampling from a population having a normal distribution, the standard deviation may be estimated by

$$\hat{\sigma} = \frac{\overline{R}}{d_2} \tag{5}$$

where

$$\overline{R} = \text{average of the subgroup ranges}$$

$$d_2 = \int_{-\infty}^{\infty} [1 - (1 - \alpha_1)^n - \alpha_1^n] dx_1$$

$$\alpha_1 = \frac{1}{\sqrt{2\pi}} \int_{-\infty}^{x_1} e^{-\left(\frac{x^2}{2}\right)} dx$$

and n is the subgroup size [ASTM, 1990, p. 92]. Values for d_2 may be found in a table in ASTM.

The moving range (MR) is a range of $n = 2$ measurements, so (5) may be rewritten as

$$\hat{\sigma} = \frac{\overline{MR}}{d_2} \tag{7}$$

If the UCL(X) is taken to be the estimate of 3σ above the overall average, then

$$UCL(X) = \overline{X} + 3\hat{\sigma}$$

$$= \overline{X} + 3\frac{\overline{MR}}{d_2} \qquad \text{using (7)}$$

$$= \overline{X} + 3\frac{\overline{MR}}{1.13}$$

since $d_2 = 1.13$ for $n = 2$

$$= \overline{X} + 2.66\overline{MR}$$

<u>Control Limits for Xbar and s with Unequal-Size Subgroups</u>

There are three accepted ways of estimating σ [SAS, 1995, pp.1214 – 1215]. Two are given here.

1. ASTM method [1990, pp. 68 – 70]:

$$\hat{\sigma} = \frac{\dfrac{s_1}{c_4[n_1]} + ... + \dfrac{s_k}{c_4[n_k]}}{k}$$

where k is the number of subgroups, s_i is the sample standard deviation of the ith subgroup, n_i is the size of the ith subgroup (all $n_i \geq 2$), and $c_4[n_i]$ is the c_4 value for n_i.

2. Minimum variance linear unbiased estimate method:

$$\hat{\sigma} = \frac{\dfrac{h_1 s_1}{c_4[n_1]} + \ldots + \dfrac{h_k s_k}{c_4[n_k]}}{h_1 + \ldots + h_k}$$

where

$$h_i = \frac{c_4^2[n_i]}{1 - c_4^2[n_i]}$$

k, s_i, and $c_4[n_i]$ defined as above.

For the s chart: For each subgroup i,

$$\bar{s}_i = \hat{\sigma}\, c_4[n_i]$$ using (1)

and

$$UCL(s_i) = B_4[n_i]\,\bar{s}$$

$$LCL(s_i) = B_3[n_i]\,\bar{s}$$

where

$B_4[n_i] = B_4$ for n_i
$B_3[n_i] = B_3$ for n_i

$$B_4[n_i] = 1 + \frac{3}{c_4[n_i]}\sqrt{1 - c_4^2[n_i]}$$

$$B_3[n_i] = 1 - \frac{3}{c_4[n_i]}\sqrt{1 - c_4^2[n_i]}$$

This implies that for unequal subgroup sizes, the centerline and the control limits will step, that is, the centerline and control limits will differ with the subgroup size.

For the Xbar chart:

$$\bar{\bar{X}} = \frac{\displaystyle\sum_{i=1}^{k}\sum_{j=1}^{n_i} X_{ij}}{\displaystyle\sum_{i=1}^{k} n_i}$$

and

$$UCL(\bar{X}_i) = \bar{\bar{X}} + A_3\,\bar{s}$$

$$LCL(\bar{X}_i) = \bar{\bar{X}} - A_3\,\bar{s}$$

where

$$A_3[n_i] = \frac{3}{c_4[n_i]\sqrt{n_i}}$$

This implies that for unequal subgroup sizes, the control limits will step (i.e., the control limits will differ with the subgroup size).

Appendix 4 Short Answers to Selected Problems

Chapter 1

1. Variables   3. Variables   5. Attribute   7. Variables   9. Attribute   11. Common

Chapter 2

1.

	a	b
Average	7.5	4.8
Range	11	5
s	4.80	2.17

5.   a. 0.05, 0.10, 0.15, . . . , 0.85, 0.90, 0.95
   b. 0.025, 0.05, 0.075, 0.10, . . . , 0.925, 0.95, 0.975

Chapter 3

1.   Yes Yes
3.   CL = 89.4. 3 "obviously excessive" points.
5.   CL = 91.6. No indications that suggest nonrandom influence.
7.   CL = 96.25. No indications that suggest nonrandom influence.

Chapter 4

1.   UCL = 165.7, CL = 101.9, LCL = 38.1. No evidence of special-cause variation. Sufficient data for trial limits.
3.   UCL = 151.1, CL = 84.4, LCL = 17.6. Points 5 and 15 are above the UCL.
5.   UCL = 173.7, CL = 94.4, LCL = 15.1. Point 22 is above the UCL.
7.   UCL = 142.4, CL = 96.2, LCL = 50.1. No points outside of 3-sigma limits, but too few points to infer a state of control. Point 15, 2 of 3 successive points in lower zone A or beyond. Point 22, 4 of 5 successive points in lower zone B or beyond.
9.   UCL = 146.8, CL = 87.0, LCL = 27.2. Too few points to infer a state of control. No indications to suggest nonrandom influence.

Chapter 5

1.

	Xbar	s
UCL	29.02	4.65
CL	25.68	2.05
LCL	22.33	0

Xbar chart is out of control.

3.

3-sigma		Xbar	s
	UCL	179.49	82.68
SURG	CL	123	39.58
	LCL	66.51	0
	UCL	158.16	51.45
ANES	CL	123	24.63
	LCL	87.84	0

All in control.

5.

3-sigma	Xbar	s
UCL	119.72	38.22
CL	98.0	22.27
LCL	76.28	6.32

Both charts in control.

7.

3-sigma	Xbar	s
UCL	37.79	4.95
CL	36.57	4.08
LCL	35.34	3.21

Both charts out of control.

Chapter 6

1.

	I Chart	Xbar $\pm 3s$	Probability Plot
UL	52.2	53.3	60
LL	-11.1 (0)	-12.1 (0)	0

Because times are nonnegative, lower limits are set to 0.

Conclusion: Use probability plot.

3.

	I Chart	Xbar $\pm 3s$	Probability Plot
UL	201.1	247.0	350
LL	9.3	-36.6 (0)	52

The probability plot is best.

Chapter 7

1. 2-sigma limits LCL = 72.8, CL = 92, UCL = 111.2. Out of control.
3. 2-sigma limits LCL = 303.36, CL = 340.25, UCL = 377.14. Same out-of-control pattern as problem 1. C chart in problem 1 was not a satisfactory approximation for the u chart because the areas of opportunity varied so greatly.
5. LCL varies, CL = 0.0078, UCL varies. No.

Chapter 8

1. CL = 0.160, LCL, UCL vary. No indications to suggest nonrandom influence.
3. CL = 0.146. Limits vary.

5. $T = 2$, $CL = 0.149$. Limits vary.
7. $T = 2$, $CL = 0.082$. Limits vary. Nonrandom influence.
9. Special-cause. Yes. No.

Chapter 9

1. Fourth root and natural logarithm.

Chapter 10

1. Xbar and s
3. Xbar and s
5. c
7. I
9. p

References

ASTM (American Society for Testing and Materials). *ASTM Manual on Presentation of Data and Control Chart Analysis,* MNL7, 6th ed. Philadelphia: American Society for Testing and Materials, 1990.

ANSI (American National Standards Institute) Standards Z1.1, Z1.2, Z1.3 (American Society for Quality Control Publications B1 and B2). "Guide for Quality Control; Control Chart Method for Analyzing Data; Control Chart Method for Controlling Quality during Production." New York: American National Standards Institute, 1958, 1975.

Benneyan, James C. "Statistical Quality Control Methods in Infection Control and Hospital Epidemiology, Parts I and II." *Infection Control Hospital Epidemiology*, vol. 19, pp. 194 – 214, 265 – 283, 1998.

Bicking, Charles. A., and Frank M. Gryna, "Process Control by Statistical Methods." Section 23 in *Quality Control Handbook*, J. M. Juran, editor-in-chief, 3rd ed.,. New York: McGraw-Hill, 1974.

Brassard, Michael. *The Memory Jogger +*. Methuen, MA: GOAL/QPC, 1996.

Clemmer, Terry P., Vicki Spuhler, Thomas Oniki, and Susan Horn. "Results of a Collaborative Quality Improvement Program on Outcomes and Costs in a Tertiary Critical Care Unit." *Critical Care Medicine*, vol. 27 (9), pp. 1768 – 1774, 1999.

Cleveland, William S. *Visualizing Data*. Summit, NJ: Hobart Press, 1993.

Deming, W. Edwards. "On Some Statistical Aids toward Economic Production." *Interfaces*, vol. 5 (4), pp. 1 – 15, August 1975.

Deming, W. Edwards. *Quality, Productivity, and Competitive Position*. Cambridge, MA: Massachusetts Institute of Technology, Center for Advanced Engineering Study, 1982.

Duncan, Acheson. *Quality Control and Industrial Statistics*, 5th ed. Homewood, IL: Richard D. Irwin Inc., 1986.

Eddy, David M. "Performance Measurement: Problems and Solutions." *Health Affairs*, pp. 7 – 25. July/August 1998.

Geary, R. C. "Testing for Normality." *Biometrika*, vol. 34, pp. 209 – 242, 1947.

Grant, Eugene, and Richard Leavenworth. *Statistical Quality Control*, 7th ed. New York: McGraw-Hill, 1996.

Hart, Marilyn. "C Control Chart Limits for a Given Number of Subgroups." *Proceedings of the Midwest Business Administration Association*, Chicago, pp. 71 – 74, March 13 – 15, 1996.

Hart, Marilyn. "R and s Control Chart Limits for a Given Number of Subgroups." *Proceedings of the Western Decision Sciences Institute*, Seattle, WA, pp. 335 – 339, April 2 – 6, 1996.

Hart, Marilyn. "NP Control Chart Limits for a Given Number of Subgroups." *Proceedings of the Decision Sciences Institute*, San Diego, vol. 2, pp. 994 – 996, November 22 – 25, 1997.

Hart, Marilyn, and Robert Hart. *Quantitative Methods for Quality and Productivity Improvement.* Milwaukee, WI: ASQC Quality Press, 1989.

Hart, Marilyn, and Robert Hart. "X-Bar Control Limits for an Arbitrary Number of Subgroups." *Proceedings of the Western Decision Sciences Institute*. Maui, Hawaii, pp. 637 – 639, March 29 – April 2, 1994.

Hart, Marilyn, Robert Hart, and George Philip. "Control Chart False Alarm Risks with Three Sigma Limits." *Quality Engineering*, vol. 4(3), pp. 413 – 418, 1992.

Hart, Robert, and Marilyn Hart. "Shewhart Control Charts for Individuals with Time-Ordered Data." *Frontiers in Statistical Quality Control*, #4, pp. 123 – 137, H.-J. Lenz, G.B. Wetherill, and P.-Th. Wilrich, editors. Heidelberg, Germany: Physica-Verlag, 1992.

Hilliard, Jim E., and H. Alan Lasater. "Type One Risks when Several Tests Are Used Together on Control Charts for Means and Ranges," *Industrial Quality Control*, pp. 56 – 61, August 1966.

Joint Commission on Accreditation for Healthcare Organizations (JCAHO). *Accreditation Standards*. Oak Brook Terrace, IL: JCAHO, 1999.

Juran, Joseph M. *Management of Inspection and Quality Control*. New York: Harper and Brothers, 1945.

Montgomery, Douglas C. *Introduction to Statistical Quality Control*, 3rd ed. New York: Wiley, 1997.

National Institute of Science and Technology. "2001 Health Care Criteria for Performance Excellence." Malcolm Baldrige National Quality Award Program. Online, available at www.quality.nist.gov, accessed May 1, 2001.

Nelson, Lloyd S. "Control Charts for Individual Measurements," *Journal of Quality Technology*, vol. 14(3), pp. 172 – 173, July 1982.

Nelson, Lloyd S. "The Shewhart Control Chart—Tests for Special Causes," *Journal of Quality Technology*, vol. 16(4), pp. 237 – 239, October 1984.

Nelson, Lloyd S. "Interpreting Shewhart Xbar Control Charts," *Journal of Quality Technology*, vol. 17(2), pp. 114 – 116, April 1985.

Newcomer, Lee N. "Physician, Measure Thyself." *Health Affairs,* pp. 32 – 35, July/August 1998.

O'Leary, Dennis S. "Reordering Performance Measurement Priorities." *Health Affairs*, pp. 38 – 39, July/August 1998.

Ott, Ellis. *Process Quality Control*. New York: McGraw-Hill, 1975.

Pagano, Marcello, and Kimberlee Gauvreau. *Principles of Biostatistics*, 2nd ed. Pacific Grove, CA: Duxbury, 2000.

Pearson, E. S. *The Application of Statistical Methods to Industrial Standardization and Quality Control*, British Standard Number 600. London: British Standard Institution, 1935.

SAS Institute Inc. *SAS/QC Software: Usage and Reference*, version 6, ed. 1, vol. 2. Cary, NC: SAS Institute Inc., 1995.

Schilling, Edward G. "A Systematic Approach to the Analysis of Means." *Journal of Quality Technology*, pt. 1, vol. 5(3), pp. 93 – 108, July 1973, and pts. 2 and 3, vol. 5(4), pp. 147 – 159, October 1973.

Shapiro, Samuel. *How to Test Normality and Other Distribution Assumptions*. Milwaukee, WI: American Society for Quality, 1990.

Shewhart, Walter A. *Economic Control of Quality of Manufactured Product*. Princeton, NJ: Van Nostrand Reinhold Co., 1931. (Republished in 1981 by the American Society for Quality Control, Milwaukee, WI).

Shewhart, Walter A. *Statistical Methods from the Viewpoint of Quality Control.* Washington, D. C: The Graduate School, Department of Agriculture, 1939.

Taguchi, Genichi. *On Line Quality Control during Production*. Tokyo: Japanese Standards Association, 1981.

Western Electric Company (now AT&T). *Statistical Quality Control Handbook*, Bonnie Small, editor. Newark, NJ: Western Electric Company, Inc., 1956.

INDEX

Joint Commission on Accreditation of Healthcare Organizations (JCAHO), 1, 4
judgment (state of control), 4, 5, 70, 97, 108, 167, 198
Juran, 7, 57

KJ diagram, 10

lack of control, 71, 74, 112, 113, 142, 160, 198
lack of statistical control, 59
Lasater, 70
Leavenworth, 58, 158, 162
length of stay (LOS), 15, 16, 17
level, 45, 46, 61, 74, 86, 87, 161, 202
limits for individuals, 94

maintenance, 128, 129
Malcolm Baldrige, 11
mean, 4, 15, 33, 38, 47, 58, 61, 83, 84, 88, 94, 105, 148, 210, 272, 304, 329
measurement data, 59, 157, 285, 289
median, 29, 48, 108, 156, 304
minimum number of plotted points, 58
minimum subgroup size, 158, 160, 162, 168, 190, 191, 192, 199, 218, 219, 222, 230, 241, 296
mixture, 302
Montgomery, 309
moving range, 61, 62, 74, 96, 331
MR chart, 62, 77

natural limits, 62, 100, 131, 132, 136, 137

near-normal, 22, 23, 27, 33, 34, 45, 59, 63, 71, 76, 90, 95, 96, 98, 100, 102, 125, 129, 138, 139, 140, 151, 243, 244, 247, 250, 251, 252, 258, 261, 262, 272, 277, 278, 279, 282, 286, 292, 294, 298
Nelson, 48, 49, 61, 62, 64, 65, 66, 67, 68, 69, 167, 198, 301, 303, 306
Newcomer, 1
no standard given, 46, 50, 57, 60, 63, 69, 70, 71, 77, 81, 88, 125, 273
normal distribution, 21, 23, 25, 30, 34, 58, 62, 95, 131, 132, 134, 136, 137, 144, 147, 292, 304, 329, 330, 331
np chart, 199
number of subgroups, 16, 59, 60, 61, 71, 159, 224, 331

ongoing process control, 70, 77, 93, 94, 98, 116, 119, 128, 160, 166, 170, 171, 172, 173, 192, 204, 206, 207, 209, 211, 212, 231, 247, 248, 250, 259, 264, 265, 273, 282, 287
Ott, 60
outputs, 2

p chart, 58, 180, 181, 189 ff, 285, 286, 287, 295, 296, 297, 302, 312, 317, 318, 319, 323
Pagano, 316
Pareto chart, 1, 6, 14
patterns, 3, 10, 45, 47, 48, 88, 93, 108, 112, 181, 302
Pearson, 3, 57
Philip, 69
Poisson, 157, 158, 162, 167, 168, 249, 285, 303

power, 45, 59, 60, 224, 237, 239, 241, 243, 244, 248, 250, 264, 265, 267, 268, 272, 279, 281, 282, 297, 311
preliminary control limits, 116, 119
probability plot, 6, 10, 27, 29, 31, 33, 34, 35, 59, 63, 71, 90, 98, 131, 132, 133, 134, 136, 137, 140, 141, 144, 147, 148, 149, 151, 153, 156, 243, 244, 248, 250, 252, 261, 262, 272, 278, 279, 282, 286, 290, 292, 298, 334
process capability, 5, 27, 33, 34, 62, 70, 94, 131, 132, 133, 134, 136, 137, 139, 140, 141, 142, 144, 147, 148, 149, 153, 156, 157
process improvement, 4, 5, 13, 47, 51, 58, 60, 63, 70, 77, 78, 88, 97, 100, 102, 105, 113, 125, 167, 183, 186, 198, 231, 238, 241, 259, 265, 267, 268, 281, 289, 295

R chart, 101, 302
random, 3, 13, 45, 48, 59, 60, 88, 108, 124, 134, 180, 202, 219, 234, 265, 266, 271, 286, 294, 297, 304, 313, 315, 317, 318, 319, 322, 324
range, 10, 16, 17, 19, 23, 33, 35, 61, 74, 101, 331
rational subgroup, 58, 101, 102, 103, 105, 106, 108, 113, 116, 119, 128, 160, 167, 186, 189, 195, 198, 204, 209, 218, 219, 227, 228, 232, 287
relative cumulative frequency, 27, 28, 34
retrospective, 46, 47, 71, 93, 282